墨香财经学术文库

U0805560

The Effect of Photo Decoration Cues on Consumers

图片装饰线索对消费者的影响

周峰 黄仕晖 张长峰 丁光亮 ◎ 著

东北财经大学出版社
Dongbei University of Finance & Economics Press
大连

图书在版编目（CIP）数据

图片装饰线索对消费者的影响 / 周峰等著. —大连：东北财经大学出版社，2018.12
（墨香财经学术文库）
ISBN 978-7-5654-3354-2

Ⅰ. 图…　Ⅱ. 周…　Ⅲ. 食品包装-影响-消费者行为论-研究　Ⅳ. ①TS206
②F713.55

中国版本图书馆CIP数据核字（2018）第257067号

东北财经大学出版社出版发行
大连市黑石礁尖山街217号　邮政编码　116025
网　　址：http：//www.dufep.cn
读者信箱：dufep @ dufe.edu.cn
大连永盛印业有限公司印刷

幅面尺寸：170mm×240mm　字数：175千字　印张：10　插页：1
2018年12月第1版　　2018年12月第1次印刷
责任编辑：张晓鹏　吉　扬　　责任校对：至　庄
封面设计：冀贵收　　版式设计：钟福建
定价：45.00元

教学支持　售后服务　联系电话：（0411）84710309

如有印装质量问题，请联系营销部：（0411）84710711

前言

当前，随着电子商务竞争日趋激烈，在线网店开始转向注重产品的视觉设计来吸引消费者的注意力。近年来，生鲜网店的图片装饰线索变得越来越普遍，图片装饰线索让消费者能够体验到购物的乐趣和视觉上的美观，从而解决了消费者因无法接触真实产品所带来的购买问题。之前的理论研究大都注重网店的设计风格、网店总体色彩使用和产品图片的效果，而针对产品图片的装饰线索并没有在管理信息系统和电商领域引起学者们的足够注意。产品图片就是对产品的一种展示，是对产品文字描述的补充，不仅能够显示产品的全貌，更能让顾客看到产品的每一个细节、做工以及产品的质感，从而让顾客感受到产品的价值，因此高品质的产品图片装饰就显得尤为重要。为此，作者在占有大量文献的基础上，提炼出三种图片装饰线索：产品反射效果、对比色的使用和互补食材的布局设计，并构建模型，提出研究假设，收集大量样本数据。通过研究分析，结果表明，图片装饰线索能对消费者的情感感知、购买态度和行为意图产生积极影响，同时又验证了媒体线索的信息传递能力能够吸引消费者对所购产品的关注。本书的特色在于首次将媒体丰富理论

（Media Richness Theory）和视觉修辞理论（Visual Rhetoric Theory）扩展到电子商务领域，丰富了管理信息系统（MIS）文献，为相关研究者提供了一种新的思路和路径。本书的相关结论在人机交互（HCI）、网店的界面设计、在线产品推广、在线广告等领域都能产生重要影响。

本书的出版获得了山东省科技厅重点研发计划项目“海上粮仓与‘一带一路’海陆冷链对接工程关键技术研发与应用”（项目编号2016GGH 4526）的资助，在此表示感谢，同时也感谢东北财经大学出版社对本书的大力支持。

本书由山东商业职业技术学院的周峰、黄仕晖、张长峰、丁光亮合著。由于水平有限，书中难免存有不妥之处，敬请广大学者批评指正！

作　者

2018 年 10 月

PREFACE

As competition in e-commerce becomes fiercer, online grocery stores are attempting to attract consumers' attention by exploiting visual design artifacts in product photos. This study refines three photo decoration cues: reflective surface, contrast color usage and complementary goods layout. These three photo decoration cues are validated in a laboratory experiment. Photo decoration in the online grocery store has been becoming prominent recently because it enables consumers to experience groceries with enjoyable and visual appeal, thereby mitigating the problems associated with consumers' absence of physical contact with products. Despite the several previous efforts that studied the effects of product photo presentations, the photo decoration cues underlying these presentation methods have not been explored in a comprehensive way in the marketing and information systems field. This study investigates a model showing how these photo decoration cues would induce consumers' affective perception, attitudes and behavioral intention.

The purpose of this study is to seek answers to those two research questions: How can photo decoration cues in online grocery be identified? Could photo decoration cues significantly have effect on the affective perception and behavioral intention of consumers? Using the media richness theory, visual rhetoric theory and visual design literature as the theoretical foundation, the study provides a solid foundation to understand the impact of photo design artifacts on consumers' affective responses, attitudes and behavioral intention when online consumers shop for fresh products.

The findings provide strong support for the proposed model, and the results indicate (1) that reflective surface, contrast color usage and complementary goods layout lead to consumers' affective reactions; (2) contrast color usage does not significantly affect consumer shopping enjoyment. The findings indicate that photo decoration cues are the important construct and act as an antecedent to affective responses. In addition, both visual appeal and shopping enjoyment will positively impact consumers' attitudes towards product and attitudes towards store, moreover, both attitudes towards product and attitudes towards store are antecedents of purchase intention. Furthermore, our findings suggest that the media cues' conveyance ability to present product and deliver information has potential to attract consumers' attention to the purchase task. Therefore, the cues' conveyance ability is more likely to attract attention compared to a media selection.

This study contributes to the information systems literature by extending the theories of media richness and rhetoric to the e-commerce context. It contributes to design practice by revealing that visual design artifacts can be more effective in attracting consumers towards the products and website. The findings have important implications for website interface design, online product promotion and online advertising, among others. In addition, from the human-computer interaction (HCI) perspective, this study contributes the research of HCI and e-commerce to understanding the impact of photo design artifacts on consumers' affective responses.

LIST OF TABLES

LIST OF FIGURES

CONTENTS

Chapter 1 General Introduction

1.1 Introduction

This book refines three decoration cues from the photo in the online grocery store which enable consumers to experience grocery with affective responses. Three cues on consumers' affective perception, attitudes as well as behavioral intention are examined in a laboratory experiment using a simulated online grocery store. The findings provide strong support for the proposed model, and the results indicate (1) both reflective surface and complementary goods layout lead to visual appeal and shopping enjoyment; (2) contrast color usage positively impacts on visual appeal while it does not significantly affect shopping enjoyment; (3) consumers' affective responses positively impact consumers' attitude towards product and attitude towards store which in turn lead to purchase intention. The finding shows that photo decoration cues acting as antecedent of affective responses have important implications for website interface design, online retailing, online product promotion and online advertising. In addition, from the human-computer interaction (HCI) perspective, this study also contributes to the research on HCI and e-commerce to understand the impacts of photo visual design artifact on consumers' affective responses. Contributions, limitations along with areas for future research are discussed.

1.2 Research Background

1.2.1 Grocery Shopping

Shopping for grocery is believed to be an energy-wasting and time-consuming progress (Aylott and Mitchell, 1998; Geuens et al., 2003) compared with other shopping activities which consumers perceive more enjoyable. A survey, conducted by Peapod.com—an online grocery store, also finds that shopping for grocery is a household chore that most consumers dislike to do just like they are unwilling to go to see the dentist (Corral, 1999). What's more, a research finding from the University of Michigan further validates the phenomenon by showing that shopping for grocery is ranked at the bottom just in front of cleaning among 22 different household activities (Richards, 1996). Hence, Datamonitor (2008) argues that consumers are eager to seek an efficient and convenient way for shopping grocery.

Verhoef and Langerak (2001) find that one notable feature of the consumers is that they prefer to purchase grocery products via their sensory channel. When consumers buy a grocery product, to a large extent, they greatly rely on sensory intuitions such as seeing, smelling, feeling or touching to judge the freshness of the product, especially for fresh food, such as fruit, vegetables, fish and meat, which possess rich characteristics in sensory attributes (Geuens et al., 2003). Researches have also identified that the sensory attributes are presented in an effective format on the website which appears to be a critical element that can decrease consumers' risk perception, and thus result in consumers' perceived compatibility (Lim et al., 2009) with physical grocery products.

1.2.2 Online Grocery Shopping

Shopping online for grocery refers to a form of electronic commerce which allows the consumers to order grocery online at home or other places and the following logistic work of those products (Ghazali et al., 2006). The very convenient and rapid development of information technology enables an increasing number of consumers to search and buy grocery products online. Online shopping for some certain grocery products is obviously different from shopping for other types of products in the store, especially for fresh foods that possess riche sensory property (Geuens et al., 2003; Morganosky and Cude, 2000). Therefore, purchasing groceries online is deemed to be a sustainable innovation procedure (Robinson et al., 2007; Hansen, 2005a). Anderson and Ortinau (1988) argue that marketing strategies should be continuously innovated to satisfy the ever changing needs of the online consumers.

With the maturity of online shoppers and intensified competition among online vendors, more and more online grocery stores are expecting to increase their sales, improve consumers' satisfaction and enhance the efficiency in delivery and service (Avery et al., 2012). Correspondently, both researchers and marketing professionals have shown a great surge of interest in exploring consumers' behaviors towards this new purchase channels.

1.2.3 Online Grocery Market

Online grocery shopping is believed to be a quite new market which develops rapidly in recent years. Online grocery market is depicted as what's the mass group eating and drinking sold online, including online-based grocers' sales as well as delivery scheme (Mintel, 2011). Online grocery market has enjoyed sustained increase with the growth rate of 14% each year in the world (Halzack, 2015), and it is predicted that this sector will account for about 10% to 12% of the market by 2020 (Rankin, 2013) in most countries. According to the report, more than 50% UK households order grocery online and many large supermarkets are extending their online grocery stores in the United States (Briggs, 2014). European merchants are developing more collection points to meet the higher needs for online grocery purchasing. Similarly, in Australia, online grocery retailing is also developed quickly based on a low level. But the trend is particularly pronounced in China. Based on the reports from Management Consultancy McKinsey (2016), 40% of Chinese consumers shop grocery online which makes China as the global leader for online grocery sales. In line with the 2014—2015 Chinese agricultural e-commerce development report, Chinese grocery product market amounted to 12 billion yuan (Chinese RMB) in 2013, up 221% from a year earlier, and the sales exceeded 26 billion yuan in 2014.

In addition, according to the top 10 leading markets of online grocery (IGD.com 2015), by 2015, online grocery sales were \$41 billion, \$15 billion and \$12 billion in China, UK and Japan respectively. China online grocery sales will reach \$178 billion, UK and Japan will reach \$28 and \$22 billion respectively by 2020 (IGD.com 2015). See Table 1.1.

Especially for China online grocery market, the share reached 41.2 billion yuan in 2015, and this number would exceed 66 billion yuan in 2016, and it is predicted that this number will surpass 100 billion yuan in 2018 (iiMedia Research, 2015). The statistical results are shown in Figure 1.1.

Attracted by this potential huge profit, an increasing number of online grocery stores will enter into this new market. Nowadays, online grocery market is becoming more and more globalized and thus is regarded as a new ocean with a huge potential and an extensive market space (Wang et al., 2012).

Table 1.1 The Top 10 Leading Online Grocery Markets' Current Value and Estimated Value for 2020

	2015 ($7bn)	2020 ($18bn)
China	41	178
UK	15	28
Japan	12	22
US	7	18
France	9	16
Republic of Korea	7	13
Germany	3	6
Australia	2	5
Belgium	1	2
Netherlands	0.5	2

Source: IGD.com, 2015

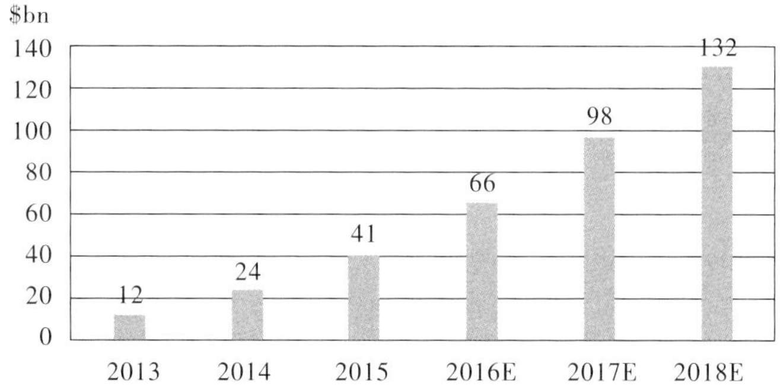

Source: iiMedia Research, 2015

Figure 1.1 China Online Grocery Market and Prediction

1.3 The Necessity of Research

Research has shown that the inadequate online product presentation restricts the development of e-commerce (Rose et al., 1999). To alleviate this concern, various photo process methods have been employed to enrich product presentations. Most previous studies associated with product presentation mainly focus on the media selection among photo, text and other interactive ways (e.g. 3D). It has been proved that media selection is not enough to present rich product information, particularly for experience attributes (Nelson, 1974). Although some researches start to investigate the effects of photo details which have played a role on the consumer reactions, the detail photo visual design artifacts (i.e. photo decoration cues) underlying these presentation methods have not been explored in a comprehensive way especially for grocery products. Therefore, it is necessary to have deep insight on the impact of photo visual design artifacts on grocery e-commerce success in grocery retailing area.

1.4 Research Objective

With the rapid development of online grocery market, undoubtedly, consumers' online grocery shopping motivation is worth investigating. However, seldom is understood how consumers shop online for grocery and what the patterns are comparing with physical store shopping behaviors. As online shopping has evoked sufficient research interest, investigating the reason why online grocery shopping booming is now just emerging. Based on the literature review, this study proposes a research model showing how grocery photo decoration cues induce consumers' affective motivation, attitudes and behavioral intention. The objective of this study is to investigate whether and how online grocery photo decoration cues exposure influences on consumers' affective perception. Consequently, this study is expected to make up an important gap in product presentation literatures on online grocery presentation.

1.5 Research Question

The use of photo in website to draw users' attention (Riegelsberger, 2002) and enhance online products' credibility has been confirmed (Fogg et al., 2002). Thus, by presenting photo visual decoration cues, online vendors can effectively influence consumers' attitudes towards product and website. A photo with decoration cues exposes consumers to more information and thus can stimulate consumers' affective responses.

However, there are limited previous studies which have been done on the effects of photo decoration cues in online grocery store on consumers' response and purchase decision-making. Towards this end, the purpose of present study is expected to make up an important gap in product presentation literatures on online grocery product presentation. Hence, we strive to answer the following research questions. (1) What photo decoration cues in online grocery could be identified? (2) How could photo decoration cues significantly affect consumers' affective perception (i.e. visual appeal and shopping enjoyment), attitude (i. e. attitude towards product and store), and behavioral intention (i.e. purchase intention)?

1.6 Research Methodology

A laboratory experiment is performed to validate the proposed research model and to test the hypotheses. Two experiments are adopted in this study. The first experiment aiming to confirm three types of photo decorative cues are confirmed and accepted by consumers which ensure that this manipulation would also be successful in the main experiment. In the main experiment, reflective surface, contrast color usage and complementary goods layout at two levels are manipulated respectively. On the post experimental questionnaire, participants are asked to rate the degree to which each item is experiential by photo decoration cues on a seven-point Likert scale. 331 valid data have been collected and analyzed using ANOVA and PLS-SEM algorithm respectively.

1.7 Research Contributions

This study contributes to the Information Systems (IS) and marketing literatures by extending the theories of media richness and visual rhetoric to the e-commerce context. It contributes to the design practice by revealing that visual design artifacts can be more effective in attracting consumers' attention towards product and online grocery store. The findings have important implications for website interface design, online grocery retailing, product presentation, product promotion, and online advertising among others. In addition, from the HCI perspective, this study contributes to the researches on HCI and e-commerce to understand the impact of photo design artifacts on consumers' affective responses and purchase decision making.

1.8 Research Outline

This book is organized as follows. Chapter 1 shows a general introduction about this research. Chapter 2 reviews the existing literature on media richness theory, visual design and the theory of rhetoric, consumers' attitude towards product and attitude towards store, and photo decoration cues are also discussed. In the following chapter, a research model is developed to validate how photo decoration cues affect consumers' behaviors through the effective responses, attitude towards grocery store and attitude towards products; along with above argument hypotheses are developed. Chapter 4 presents the research methodology and introduces laboratory experiment procedure. Chapter 5 analyzes the data to test the research model and hypotheses. Conclusion, implications, contribution, research limitations and future research direction are discussed in Chapter 6. Survey questionnaire, experiment website, instrument and some terms used in this research are also listed in the appendices respectively.

Chapter 2 Literature Review

2.1 Photo Decoration Cues

As shopping online is sweeping across the world, both online grocery stores and consumers have been concerned about how product is presented as well as processed by customers (Burke, 2002). For example, research finds that the detailed information of product presentation significantly influences consumers' satisfaction (Szymanski and Hise, 2000). Jarvenpaa and Todd (1997) also argue that understanding of product is one of the essential elements influencing consumers' attitudes towards product and purchase behavior. For this reason, both researchers and practitioners strive to improve online grocery product presentations.

Research has shown that the inadequate online product presentation is a main obstacle for e-business (Rose et al., 1999). In offline context, consumers can directly access the groceries as well as inspect them in order to make purchase decisions. In contrast, in the online shopping environment consumers are unable to physically examine the product prior to purchase, hence, more detailed information is needed for consumers to lower their decision making uncertainty (Laroche et al., 2005).

To alleviate this concern, various photo process technologies have been adopted currently to enrich product presentations in most grocery stores. Photo is one type of important media for consumers to understand a product's quality and design. Similar to the advertising, photos can lower consumers' uncertainty about the products (Shavitt et al., 1998). It is popular for most online stores to employ photos to present products' information which benefit consumers to generate a clear comprehension of products (Hoch and Deighton, 1989) and decrease consumers' uncertainty about the products (Shavitt et al., 1998). Product photos are usually applied to describe visual appeal of products (Lightner and Eastman, 2002), which is generally difficult to be depicted by verbal cues alone (Baggett, 1989).

With the progress in photo technologies, human-computer interface designers and Information Systems researchers began to concentrate on the adoption of photos in sustaining human decision making from 1980s (DeSanctis, 1984; Jarvenpaa and Dickson, 1988). Product photos play a pivotal function in provoking users' affective and cognitive reactions, which in turn influences shopping experiences and approach behaviors (Yoo et al., 2014). For many years, photos in advertising have been adopted to show product information and to provoke affective reactions in consumers (Cry et al, , 2009). Larkin and Simon (1987) argue that "a photo is sometimes worth ten thousand words". Such that, many commercial website currently put a lot of effort into polishing photo in order to fascinate consumers' understanding product (Li et al., 2014). In order to attract consumers, a larger amount of online vendors adopt different methods of product presentations to advertise their products.

However, most previous studies associated with product presentation mainly focus on the media selection among photo, text and other interactive way (e.g. 3D) to present their products. For example, Chau et al. (2000) analyze and compare photos and text and find that when consumers order familiar products, photos surpass the text in both effectiveness and efficiency. Lightner and Eastman (2002) identify that the words on a website result in higher satisfaction than photos for product descriptions in a remote purchase environment. Suh and Lee (2005) study the impact of photos and 3D product presentation on knowledge, attitude and purchase intention, and they insist that 3D product presentation could enable customers to recall more product knowledge and enhance their purchase trend. Hassanein and Head (2007) find that product photo together with its described text can raise consumers' social presence resulting in their perceived enjoyment and trustful intention towards online shopping. Similarly, Park et al. (2008) argue that 3D products' rotation can affect consumers' emotion, perception of information, attitudes and purchase tendency. Also, both Koh and Sundar (2010) and Ou and Sia (2010) find that the balanced integration of photos and text enables consumer to generate feelings and expectations towards a concrete webpage. Other relative researches also give various findings from different perspectives towards products' photo comparing to other media (Table 2.1).

And yet, based on above research route, media selection between photo and other media is not enough to present rich product information, especially for grocery products, because it narrowly explains the detailed information in product photos (Li et al., 2014). As competition in e-commerce becomes fiercer, online grocery stores are attempting to attract consumers' attention by exploiting visual design artifacts in product photos. In the online grocery store, consumers just catch sight of some interfaces based on 2D or 3D photos.

Nowadays, many researches start to investigate the effects of photo details which have played a role on the consumer reactions. For instance, Song and

Table 2.1 Studies Based on Media Selection

Source	Findings
Chau et al. (2000)	When consumers order familiar products, photos surpass the text in both effectiveness and efficiency.
Lightner and Eastman (2002)	Experiment results identify that the words on a website result in higher satisfaction than photos for product descriptions in a remote purchase environment.
Suh and Lee (2005)	3D product presentation could help customers recall more product knowledge and increase their purchase predisposition.
Hassanein and Head (2007)	The text and photo on behalf of social presence related to product photos can raise consumers' enjoyable and trustful intention towards online shopping.
Park et al. (2008)	3D products' rotation can affect consumers' emotion, perception of information, attitudes and purchase tendency.
Jiang and Benbasat (2007)	The usage of videos or virtual interaction technology is beneficial to understand the product and can improve product knowledge.
Lin et al. (2013)	The ratio of photo to text affects users' feelings towards a webpage.
Verhagen et al. (2014)	The virtual mirror enables to create local presence and generate product likelihood and product tangibility.
Koh and Sundar (2010) Ou and Sia (2010)	The balanced integration of photos and text enables consumers to generate feelings and expectations towards a concrete webpage.

Kim (2012) examine the impacts of photo size on buying tendency and find that larger photos enable customers to gather more information and hence elevate their purchase tendency. Lavie et al. (2011) argue that the photos' warm colors are arousing and exciting and are usually connected with consumers' positive mood. Li et al. (2014) find that the photo with a warmer color and more social presences can well influence consumers' purchase intention. Hong et al. (2004) investigate that the usage of flash highlights on product information can inspire users' concentration. Nixon and Aguado (2012) have greatly emphasized on the efforts of automatic image-processing technology and extended high-dimensional characteristics to discern the semantics of photos. Yoo and Kim (2012) find that complementary garment items should be coordinated with the specific webpage to create favorable approach behaviors for consumers. The relative researches and their findings are listed in Table 2.2.

Table 2.2 Studies Based on Photo Characteristics

Source	Findings
Hong et al. (2004)	Usage of flash highlights on product information can inspire users' concentration.
Song and Kim (2012)	The larger photos enable customers to gather more information and hence elevate their purchase tendency.
Nixon and Aguado (2012)	Emphasize on the efforts of automatic image-processing technology and extended high-dimensional characteristics to discern the semantics of photos.
Lavie et al. (2011), Terwogt and Hoeksma (1995)	The warm colors are arousing and exciting and are usually connected with positive mood.
Li et al. (2014)	Photos with a larger object can decrease the key objects' entropy. Warmer color, higher contrast usage and more social presences can influence consumers' purchase intention.
Yoo and Kim (2012)	Complementary garment items should be coordinated with the specific webpage to create favorable approach behaviors.

Based on this research mainstream, grocery product photos' visual design characteristics have started to attract some researches' concern.

Despite several those efforts to study the effects of product photo on consumers' responses, the photo specific characteristics of visual design underlying these presentation methods have not been explored in a comprehensive way in the marketing and Information Systems fields.

In an attempt to further increase and optimize customers experiences with grocery products, many online grocery stores have made it possible for customers to understand product via products' photo decoration cues. For example, at Tmall's online grocery store, the largest B2C website of China (www.tmall.com), online customers can find some photos that highlight the grocery's brightness and reflect lights by presenting products' reflective surface which easily attract consumers to imagine that the grocery products are fresh. Some photos apply the contrast color between target product and background which enable to grab consumers' attention on the target product. Some other photos allocate various complementary ingredients as the background layout of target product which can inspire consumers' reaction and judgments, thus resulting in consumer recall the way of consuming the target product. Visual design-based grocery presentations are shown in Figure 2.1.

Figure 2.1 Visual Design-Based Grocery Presentations

Experiencing those vivid photo performances with decoration cues, customers' imagination and affective reactions can be attracted and stimulated. Like other types of media, photos could also convey different decoration cues to attract consumers. In those photos, retailers may adopt many design tactics in detail to make these pictures attractive and enjoyable for the purpose to advertise their products. This kind of online photo presentation could stimulate customers' imagination and affective reactions.

However, whether and how online grocery photo decoration cues expose influences on consumers' affective reactions has attracted seldom research concern. Towards this end, the purpose of present study is expected to make up an important gap in product presentation literatures on online grocery presentations. Hence, this study aims to understand how photo decoration cues impact consumers' affective perception, attitudes and approach behavior via groceries' photo decoration cues.

2.2 Media Richness Theory

Generally speaking, the media richness theory argues that media differs in the capacity to accelerate changes in understanding among communicators (Daft and Lengel, 1986), and media richness enables to maximize the knowledge of worker's task performance, because it can lead to effective communication (Kahai and Cooper, 2003) and negotiation success (Jiang and Benbasat, 2007b). A rich media can facilitate insight and prompt understanding (Daft et al., 1987).

The impact of media richness on the variety of cognitive, affective and conative reactions of knowledge worker has been validated in both computer-mediated communication and online shopping contexts. They find that the media richness could successfully enhance message clarity, social perception, evaluation of others and decision quality in computer-mediated communication (Kahai and Cooper 2003). They also find out that it could enhance product knowledge, perceived diagnosticity, compatibility, shopping enjoyment, attitudes towards both product and store, and consequent purchase intention (Suh and Lee 2005; Jiang and Benbasat, 2007b).

For instance, when someone says "Great job!" to others, originally, these words mean that a person's working performance is good. Reversely, assuming if this message goes with some intonation and accent, the meaning of these characters dramatically changes and it indicates the bad performance.

In this case, intonation is one cue, with the message itself it can convey different meanings in communication conditions. From this example, we can see that the cues' conveyance ability could be regarded as the essential element of media richness. Despite of this aspect, recent researches still mainly focus on the sensor breadth, sensory depth and media difference, they fail to focus on the cues' conveyance ability. By these considerations, we try to focus on the cues' conveyance ability in the grocery photos.

As mentioned above, It could be more comprehensive that one determinant of media richness could be the "cues" being provided through media to understand information effectively (Dennis and Kinney, 1998). Some determinants of media richness as sensory depth, sensory breadth and interactivity of the multimedia have been emphasized in recent studies, which commonly seem to emphasize multiple sources of sensor (Suh and Lee, 2005; Jiang and Benbasat, 2007b). However, they still mean the ability of media which could convey a variety of cues. Wider range of determinants of media richness as multiple cues, immediate feedback, and personalization, language variety (Kahai and Cooper, 2003) could also be considered in the body of media richness literature.

Jiang and Benbasat (2005) announce the critical roles of multimedia in product presentations, they insist that a vivid photo presentation can provide users with more information and provoke more stimulus than powerless photo presentations. Lim et al. (2000a) verify two features of multimedia: rich language (language itself) and "complementary cues" and declare that multimedia can integrate symbolic and processing abilities of different medias and consequently form "a richer symbolic system of communication" (p. 118). The different information cues do not promote mutual competition but complement each other and emphasize the holistic effects of the information conveyed (Lim et al., 2000a).

By these reasons, Kahai and Cooper (2003) emphasize the concept of "cues" which could be some metaphors to remind communicators along with the main body of message itself. By combining the concept of multiple cues (e.g., Kahai and Cooper (2003) define media richness as the employment of multiple channels, such as intention variation and body posture including oral message; Dennis and Kinney (1998) conceptualize it as the ability of the medium to transmit multiple cues.) and language variety (Kahai and Cooper, 2003), cues conveying ability of media could be regarded as the essential element of media richness.

Substantial previous studies have shown that proper use of multiple cues can effectively convey nonverbal information, enhance the perception and be helpful for forming impression of products (Mayer et al., 2003). To this end, photo decoration cues, a new construct is created in this study, which represents the abilities to convey grocery products' information, and consequently benefit consumer to have an insight into and assess the performance of online grocery products.

Based on the media richness theory, this study focuses on the cues' conveyance abilities in the grocery photo, and three decoration cues are adopted to convey grocery information and such that enable consumer to insight into the performance of online grocery.

2.3 Visual Design Artifacts

The multi-dimensional structure is the remarkable characteristic of the website (Kim and Stoel, 2004) and many studies classify website design structures into different categories (Tarafdar and Zhang, 2006; Tan et al., 2009; Zhang et al., 2001). Usually, from the usability perspective, three popular design structures are adopted in most studies, such as information design, visual design and navigation design (Garrett, 2003), and they are tested as antecedents to online customer reactions as trust and satisfaction (Cyr, 2009b). See Table 2.3.

Table 2.3 Website Design Dimensions

Concept	Definition	Description
ID	The information which includes accurate or inaccurate elements about products or services is exposed to user (Garrett, 2003).	Customers who are discontented with website information will abandon purchase and choose to leave the site (McKinney et al., 2002).
ND	Navigational design scheme is adopted to facilitate or dampen user to access different regions of a website (Garrett, 2003).	No matter how complete the information is, customers may leave the website as they find there is difficulty to search or get the information they need (McKinney et al., 2002).
VD	Visual design elements focus on aesthetics, balance, uniformity, emotional appeal and website ' s ensemble graphical appearances (Cyr, 2008).	Visual design core compositions include photographs, colors, shapes, silhouette and pattern or font (Garrett, 2003).

Note: 'ID' represents information design. 'ND' represents navigation design. 'VD' represents visual design.

The core compositions of visual design include photographs, colors, shapes, silhouette, pattern or fonts (Garrett, 2003), visual design artifacts as an e-commerce website characteristic is typically graphic or image-based (Garrett, 2003; Cyr, 2009b) which try to adopt lively visual effects to strengthen users' insights of product performance (Jiang and Benbasat, 2007a).

Visual design can convey to consumers a first impression via graphical and structural elements (Karimov et al., 2011). Demangeot and Broderick (2010) depict visual design as the attention drawed, aesthetic and webpages' visual quality. Tan et al. (2009) validate that website visual design elements, such as "Color usage", "Layout / Space usage", "Graphics usage" and "Information presentation" are the most effective properties of websites. Visual design elements focus on aesthetics, balance, uniformity and emotional appeal of the website holistic graphical appearances (Cyr, 2008).

Along with navigation design and information design, visual design can play an essential role to deliver some meaningful visual information in website. Inasmuch as photos are essential contents of visual design and online grocery store is one type of website, therefore, online grocery store's photo visual design artifacts might have a significant effect on consumers' perception, attitudes and purchase intention, and this study focuses on the visual design cues in grocery photos.

2.4 Visual Rhetoric Theory

Rhetoric is one category of interpretive theory which aims to enhance expressive force and persuasion that build the message as an interest party tries to impact audiences (Burke 1969; Corbett 1965). Under this metaphor condition, complex visual communication is easily delivered to the viewer via factors such as photos which carry meanings to the audience. Based on the visual rhetoric theory, it is expected that the elements of focus, graphics, visual viewpoint and layout to be interrelated in a peculiar way to the message itself (Scott, 1994, p. 255).

Visual rhetoric theorists argue that it must not be understood by advertisement photos just as the mere copy of reality, but must be across as reflections (or contortion) of reality, therefore, they can be deemed as an emblem system adopted for the purpose of persuasion (Scott, 1994). In line with Scott (1994), if construing an advertising photo as a kind of rhetoric, then visual artifacts per se must provide some kind of abilities and features. Three levels of visual rhetoric are identified as follows.

2.4.1 Inventory

The first level is inventory. Inventory represents symbols and meanings of artifacts in a picture which could be understood by consumers based on their background knowledge about an issue (Scott, 1994). In this study, a new construct for reflective surface is created as well as validated to examine this inference. In online grocery photos, reflective surface will appear when photo designers put some light effects on the target grocery. According to Scott (1994), visual artifacts in the photo must have the abilities to reflect abstractions, ideas, metaphors, actions and modifiers, thus they can be employed as a complex argument in the concept of "invention". In this condition, reflective surface represents the symbol of freshness, when consumers find the reflective surface comes from product, it can easily make the consumers feel that the target grocery must be fresh. Reflective surface can assist consumers to understand and evaluate the online grocery products' performance and quality.

2.4.2 Manner

The second level is manner. Here, manner refers to the implicit signals of brightness and color in a picture which could construct the mood and feelings and could make consumer be inspired by the atmosphere of a picture (Scott, 1994). In online grocery store, fresh products are generally displayed to coordinate with some background items. For example, we usually see a piece of meat arranged over some green vegetable, and the significant contrast color between the meat and green vegetable can easily make consumer to target at the focus product. Visual artifacts must also convey various implications along with their manner of "delivery", thus the style choice can be suggested as an intended evaluation (Scott, 1994). Similarly, this study borrows contrast color usage on behalf of manner to convey an implicit signal for attracting consumers' attention on the target product.

2.4.3 Arrangement

The third level is arrangement. It accounts for the positioning of artifacts in a picture which can convey some metaphors and suggestions for attracting consumers (Scott, 1994). Layout is one of photos design artifacts, when groceries are placed in the picture based on some matched ingredients as their background, this kind of special layout can easily provoke consumers to value products as well as recall the way of consuming them. Scott (1994) argues that the ability must be provided to direct the order of argumentation through the "arrangement" of the visual factors. Consequently, in this study, "arrangement" is replaced with complementary goods layout which connotes the products' location in the photo and its stimulated effects on consumers' reaction. The artifacts' layout can either accelerate or dampen users' navigation (Madhavaram and Laverie, 2004).

In a visual context, photos can express a wide range of affective reactions and having emotional consequences (Scott, 1994, p.269), the visual rhetoric adopted in the photo has mainly been employed to make advertising in the marketing area and it also has associated with the representations on consumer website (Salinas, 2002).

2.5 Conceptualization of Photo Decoration Cues

In an offline environment, product visual presentation is influenced by not only product design (Hollins and Pugh, 1990; Creusen et al., 2010) but also store "decoration" and product "arrangements" (Vieira, 2010). Similarly, in online environment, the store "decoration" (i.e., website design, Karimov et al., 2011; Lohse and Spiller, 1998) could effect on website quality, perceived visual appeal, perceived product quality and purchase tendency (Wells et al., 2011). Therefore, vendors could leverage some specific decorative information to facilitate consumer to understand and evaluate their products (Pavlou et al., 2001).

Based on the above considerations and the concepts of store decoration, this study creates a new concept of "photo decoration cues" used to represent groceries' detailed information. Photo Decoration Cues mean that various photographic and editing techniques (i.e. camera control, focusing, lightning, exposure, rendering and balancing) adopted to polish photo, and they intend to emphasize or hide a part or overall target contents for effective communication by changing the elements of brightness, color, graphics and layout. The concept of photo decoration cues is described in Table 2.4.

Table 2.4 Conceptualization of Photo Decoration Cues

Concept	Description	Operationalization
Photo Decoration Cues	Various photographic and editing techniques (i.e. camera control, focusing, lightning, exposure, rendering, and balancing) could be adoptable to polish pictures, and they intend to emphasize or hide a part or overall target contents for effective communication by changing brightness, color, graphics and layout.	There are multiple cues in a product photo, which are intentionally made to emphasize the strengths or to hide the weaknesses of a target product in part or the whole to attract consumers.

Take a salmon photo of online grocery store as an example, see Figure 2.2. When customers find the reflective surface of salmon, it could be a cue to make consumer think that this salmon would be fresh. Usually, photographer could intentionally adopt gradient lighting and close-up technique to polish the photo. Contrast color between the salmon and background could easily make customer focus on the picture. The photographer could display a target product with background using the contrast color intentionally. Combination of the target product and background complementary goods could easily make a customer value them as well as recall the way of consuming them. The photographer could intentionally select the complementary goods as a background in a picture.

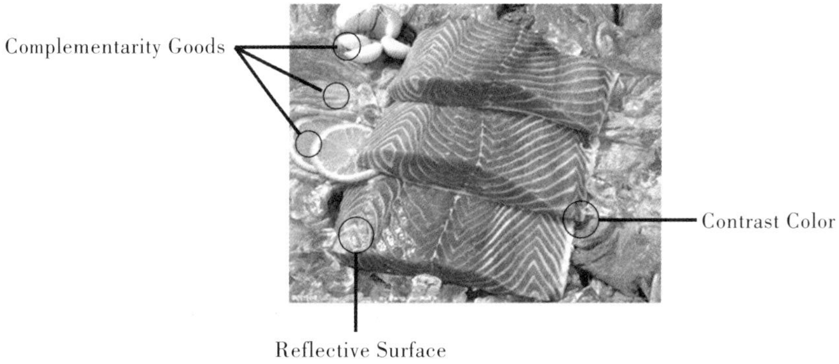

Figure 2.2 Photo Decoration Cues Demonstration in Salmon

Based on the above discussion, online grocery presentations should include multiple visual decoration cues to facilitate consumers' product understanding. The independent variables of this study are reflective surface, contrast color usage and complementary goods layout.

2.6 Consumer's Attitude towards Online Grocery Shopping

Many previous researches have proved that online grocery shopping can be accepted and understood by consumers based on their perception (Mintel 1999; Christensen and Tedlow, 2000; Eastlick and Lotz 1999; Morganosky and Cude, 2000). Consumers' attitude towards online grocery (Morganosky and Cude, 2000; Raijas, 2002) and purchase intentions for online grocery have also been examined by Verhoef and Langerak (2001). Research has shown that consumer's attitude and perception can be used to predict their behavior (Hui and Wan, 2009). Put another way, an individual's perception of online grocery shopping contributes them to perform an actual purchase for online grocery shopping. In addition, different theories and models are considered as the underlying framework has been employed to gain insight into consumers' online grocery shopping behaviors, such as the Technology Acceptance Model (TAM) and the Theory of Planned Behavior (TPB). TAM as a useful model is widely adopted to predict users' intention when applying a specific Information System (IS) in various studies (Hui and Wan, 2009). For instance, research has explored consumers' perception of online grocery shopping for the purpose of identifying different factors that affect the acceptance using TAM (Kurnia and Chien, 2003). The findings show that perceived usefulness (PU) and perceived ease of use (PEU) have positive impacts on consumers' attitudes towards online grocery product and online grocery shopping. Correspondingly, consumers' attitudes impact their behavioral intention and actual usage.

The Theory of Planned Behavior (TPB) is one of the most important social psychology theories which focus on the relationships between attitude and behavior. Based on TPB, Ajzen (1985) argues that the human behaviors are directed by three important factors: behavioral beliefs, normative beliefs and control beliefs. Among them, behavioral beliefs generate the favorable or unfavorable attitude towards the behavior. TPB is tested by Ramus and Nielsen (2005) in online grocery shopping context. Hansen (2004) has also examined that TPB is reasonable to forecast and investigate consumers' purchase intention for online grocery shopping. Some studies related to consumers' perceptions, attitudes and behavior intention towards online grocery shopping are summarized in Table 2.5. Overall, based on the TPB, in this research context, consumers' attitudes being stimulated by affective reactions can lead to the final purchase intention.

Table 2.5 Studies on Consumers' Perceptions, Attitudes and Behavior

Source	Objectives
Alba et al. (1997)	Acceptance of online shopping for consumers, vendors and product makers.
Eastlick and Lotz (1999)	Identifying individual's features, shopping patterns and consumers' attitudes towards the innovation of online shopping context.
Degeratu et al. (2000)	Comparing consumers' choice behavior in online context to the physical shopping.
Hansen (2005a)	Consumers' accepting online grocery shopping has been validated as a kind of innovation.
Hansen (2006)	What are the main factors in determining consumers' repurchase behavior towards online grocery shopping.
Hansen et al. (2004)	To validate the functions of TRA and TPB in predicting consumer online grocery purchase intention.
Hui and Wan (2009)	To examine the effects of online grocery shopping services (TAM) on consumers' purchase intention.
Kurnia and Chien (2003)	To test consumers' perception of online grocery shopping (TAM).
Ramus and Nielsen (2005)	To identify consumers' intentions for shopping grocery (TPB) when consumers shop online or offline.
Riley et al. (2007)	To validate the situational variables' function in the acceptation process of online grocery shopping
Shih (2004)	To examine consumers' acceptance of online shopping (TRA and TAM)

2.7 Consumer's Attitude towards Product and Grocery Store

It's also necessary to validate individual's attitude towards online grocery product and online grocery store to explore consumers' further behavioral intention.

Attitude towards products means consumers' overall evaluations of products, and most research works associated with consumers' purchase behavior have supposed that purchase intention is the direct outcome of consumers' attitudinal assessments towards product. The attitude towards product refers to consumers' holistic assessments about product and the likelihood intention to complete purchases on a particular website (Jiang and Benbasat, 2007b). Usually, positive attitude towards a product results in higher intention to purchase just because consumers can realize the benefits of consuming the product (Fishbein and Ajzen, 1975). Jiang and Benbasat (2007b) suggest that consumers' attitude towards product positively impacts their intention to purchase from a website. Attitude towards store refers to consumers' general evaluations of a shopping experience towards a specific online grocery store.

2.8 Summary

This section begins with a literature review about photo decoration cues. The purpose of this dissertation is to suggest that the grocery photo decoration cues should stimulate consumers' affective responses and thus lead to their affective reaction.

In order to identify photo decoration cues, understanding their role of stimulus in consumers' online grocery shopping is very necessary. It is for this purpose that this section discusses product photo, product presentation, visual design literature, media richness theory and visual rhetoric theory. Three constructs for grocery photo decoration cues are conceptualized as well as how they might provoke affective responses is explained. The three photo decoration cues are: reflective surface, contrast color usage and complementary goods layout. The findings of this study will enrich online product presentation literature and will serve as a reference for vendors to improve their grocery website design. It also has some important implications for website interface design, online product promotion, online advertising and among others. Overall, a fundamental theoretical background is provided in this section, which can better support the key constructs that will be centered on the research model and hypotheses in the following sections.

In this study, partial least squares structural equation modeling (PLS-SEM) is adopted to better understand the effects of photo decoration cues on consumers' purchase intention via several mediating variables, such as attitude towards product and attitude towards grocery store. As one of the advantages of partial least squares using the multivariate analysis technique, which employs the simultaneously analysis strategy for predictor variables and various criterions (Barclay et al., 1995), PLS approach overcomes some problems associated with correlation analyses and appropriates for explorative analysis towards structural equation models.

Chapter 3 Hypotheses Development and Research Model

Based on the literature review, this chapter demonstrates the hypotheses followed by a proposed research model.

3.1 Hypotheses Development

Visual design factors are believed to enhance the website's aesthetic playfulness and thus impact consumers' affective responses (Liu et al., 2001). In the past years, it has been examined that photos in advertising can deliver product information as well as elicit affective responses to consumers (Cry and Pan, 2009). Visual design artifacts as an important component of photos can provide detailed information about the product and can have an obvious website impact on consumers' affective responses (Eroglu et al., 2003).

3.1.1 Effects of Photo Decoration Cues on Visual Appeal

Empirical researches suggest that well-presented products provide positive affective reaction and judgment which further lead to purchasing behavior in the physical store (Turley and Milliman, 2000). Similarly, in online store context, effective and efficient product presentation not only draws consumers to a website, but also accelerates the purchase decision making of consumers who lack direct product experiences (Yoo and Kim, 2014). Online grocery vendors usually try to employ some lively visual effects to provoke consumers to understand how a grocery performs, such as visual design artifacts. Visual design artifacts can afford sufficient information to the website and photos and influence consumers' affective reactions as well (Eroglu et al., 2003).

Visual appeal can be defined as "the tangible aspect of the online environment which reflects the 'look and feel' or perceived attractiveness of a website" (Weiss et al., 2003). Visual appeal connotes the attractiveness of the website, consisting of photos, colors, and fonts (Loiacono, 2003; 2007) which are inspired by the design, beauty inherent and physical attractive appearance in the retail setting (Holbrook, 1994).

Visual search research theory suggests that the salience of visual objects can effortlessly evoke users' attention. Taylor and Thompson (1982) argue that salience means one's focus is differentially directed to the parts of environment. Salience can be endowed by local contrast among the visual characteristics such as color or layout. The salient features draw human attention and shorten the search time, which have been proved in various earlier studies.

As mentioned above, the reflective surface holds a highly salient property as it reflects the appearance of the visual surface by presenting the product with brightness and reflective light, and the photographer intentionally adopts gradient lighting and close-up technique to polish the product in order to attract consumers' visual attention.

When the reflective surface is employed in the fresh grocery product, a kind of salience comes into being and draws consumers' attention to the target product immediately. According to Scott (1994), reflective surface represents the symbol of freshness and it can easily attract consumer to feel that the grocery could be fresh. It is possible that consumers would evaluate an online grocery photo with the reflective surface to be more visually appealing. It is expected that the reflective surface can attract consumers' visual appeal towards the grocery. Therefore, we hypothesize that:

H1: Reflective surface positively affects online consumers' visual appeal.

Visual design has the potential to elicit affective appeal or behaviors for consumers (Garrett, 2003) and may be conveyed via photos or colors (Rosen and Purinton, 2004). Color as one design element has a particularly strong impact on visual appeal (Knutson, 1998) and is known to be a strong predictor of a website's overall appeal (Lindgaard 1999; Brady and Phillips, 2003). Mathwick et al. (2001) argue that the adoption of color and image quality band together influence visual appeal. Lindgaard et al. (2006) also suggest that certain fundamental aesthetic properties, such as color, have been found reliable in predicting overall visual appeal ratings across users. For instance, the majority of all sweets are visually appealing by their colors or shapes.

The concept of color contrast depicted as the change in color appearance revolves around a test stimulus material's chromaticity, including color value, color chroma and hue (Pereverzeva and Teller, 2009). Using color contrast on webpage is usually conductive to decrease users' information processing load as well as to highlight the products' presentation (Budimir and Palmovic, 2011), thus potential to shorten and simplify the road from stimulus to visual appeal. In this study, contrast color usage between the target product and background could easily attract customer to the target product. Therefore,

H2: Contrast color usage positively affects online consumers' visual appeal.

Layout is important and would portray a strong design concept and can create a deeper impression on consumer (Scapin and Bastien, 1997). Mahon-Haft and Dillman (2010) suggest that in screen design, with a proper focus on the main element, the balance between the main element and secondary elements can create balanced layout and add to visual harmony and visual comfort. Goods layout refers to the design of space, allocation of items and grouping of products. In the grocery product's photo, complementary goods layout conveys the information more imaginative and accessible by their holistic coordination effect than separating them (Tan et al., 2009). The products that are laid out vivid and aesthetic will foster consumers' concern, enjoyment, pleasure and comfort with the experiences in exploring the product (Tan et al., 2009).

In addition, in line with the theory of visual rhetoric, the arrangement accounts for the positioning of artifacts in a photo which can convey some metaphors and suggestions for attracting consumers (Scott, 1994). When groceries are surrounded by some complementary goods in the photo, this kind of layout could easily stimulate consumers' reaction and judgments. Therefore,

H3: Complementary goods layout positively affect online consumers' visual appeal.

3.1.2 Effects of Photo Decoration Cues on Shopping Enjoyment

Shopping enjoyment, as an affective assessment criterion of the shopping experience, refers to the shopping experience full of enjoyment in its own right, excluding the received product's value (Cai et al., 2009). Jarvenpaa and Todd (1996—1997) also argue that shopping enjoyment is a very critical factor accompanying consumers' online shopping experience.

Visual artifacts are deemed to enhance website's aesthetic appreciation and thus induce users' reactions (Liu et al., 2001). Research finds that design aesthetics in website have positive impact on consumers' enjoyment (Cyr et al., 2006). Multimedia presents product information through multiple cues and channels (Lim et al., 2000), which jointly form a rich and attractive information presentation and thereby can attract consumers' attention. Jiang and Benbasat (2007b) also put forward that products with more information cues in presentation contribute to the enhancement of consumers' shopping enjoyment.

According to Scott (1994), visual artifacts in the photo must have the abilities to reflect abstractions, ideas, metaphors, actions and modifiers. Reflective surface is intentionally made by a photographer with some gradient lighting and close-up technique to present products' brightness and lights. In this study, reflective surface represents the symbol of freshness, which can easily make consumers recognize that the target grocery is fresh. When customers find the reflective surface of grocery, it could be a cue stimulus to make consumers enjoy and believe that the grocery must be fresh. Complementary goods layout connotes the products' location in the photo and its stimulated effects on consumers' reaction. Layout is one of photos' design artifacts, when groceries are placed in a photo based on some matched ingredients as their background, it may draw consumer's attraction and perception about the target product. This kind of special layout can easily provoke consumers' interest to evaluate products.

In addition, empirical evidence indicates that the usage of appropriate colors could enhance consumers' pleasant feelings in e-store (Wu et al., 2008). Using color contrast on webpage is usually conductive to reducing consumers' information processing load as well as to highlighting the products' presentation (Budimir and Palmovic, 2011). Thus, it is potential to shorten and simplify the road from the stimulus to pleasantness, which in turn triggers impulse buying. Lo and Lin (2013) find that webpage's color contrast has a positive impact on consumers' enjoyable experience and propose that the online store's layout design be expected to make shopping enjoyable.

The vivid product presentation with more information cues can portray more concrete product attributes (Nisbett and Ross, 1980). When consumers' enhanced attention is combined with vivid multimedia product presentations, the feeling of enjoyment is enhanced (Webster and Ho, 1997).

From another point of view, when examining the grocery products with reflective surface, consumers are provoked to think that the target grocery must be fresh; contrast color usage can draw consumers' concentration on the target product; and complementary goods layout can stimulate consumer to imagine as well as to recall the way of consuming them. During the whole examining process, all those activities can help consumer narrow their attention down to the tasks (to examine the target grocery) and consequently make them involved in information process. According to Hoffman and Novak (1996), as a user concentrates on a specific activity, he or she loses self-consciousness and is involved in the task which induces a higher flow state. Studies on shopping enjoyment have shown that users' affective response has been investigated frequently in flow literatures, and those studies verify shopping enjoyment is an important affective component (Koufaris, 2002; Kamis et al., 2008; Van der Heijden, 2004; Sun and Zhang, 2008; Xu et al., 2013; Novak et al., 2000). For example, the consumers' emotional cues, such as color, are effectively captured by perceived enjoyment (Parboteeah et al., 2009). Applying this to the context of the present research, I propose that consumers engaging in the flow state should be demonstrated to improve intrinsic motivation and enjoyment (Tung et al.,

2006) . Indeed, research finds that responses to stimulation in shopping environment are positively related to enjoyment (Guido et al., 2007) .

In addition, the sensory stimuli supplied by a vivid presentation may also trigger intense affective responses (e.g., pleasure and arousal) that can overwhelm consumers' self-observation during online shopping (LaRose, 2001) .

In conclusion, it would be proper to expect that consumers evaluating grocery photo with reflective surface, contrast color usage and complementary goods layout will have a more pleasant flow experience which can increase their enjoyment, concentration, pleasant feelings and deep attraction towards grocery photo. Therefore, based on the above discussion, we hypothesize that,

H4: Reflective surface positively affects online consumers' shopping enjoyment.

H5: Contrast color usage positively affects online consumers' shopping enjoyment.

H6: Complementary goods layout positively affects online consumers' shopping enjoyment.

3.1.3 Affective Responses: Attitudes towards Product and Grocery Store

Even small variation in a photo can have an impact on product evaluations. For example, a slight change to the angle of camera to an advertised product will influence the attitudes of viewers (Meyers-Levy and Peracchio, 1992) . In line with Ajzen and Fishbei (1980) , attitude is depicted as an individual ' s favorable or unfavorable evaluations about performing a behavior. Attitude towards products means consumers' overall evaluation of products. Photo decoration cues can direct consumers' attention on the grocery product and consumers' attitudes may be more positive towards a product.

Consumers' evaluation about the product can be influenced by the aesthetics of various product presentations (Li et al., 2014). Visual appeal is related to the demonstration of visual design elements such as artifacts or graphics, which can enhance the overall presentation of a website (Parboteeah et al., 2009). Most previous studies show that visual appeal can indirectly influence consumers' impulse purchase, mediated by other variables (Wells et al., 2011), such as satisfaction towards the product. Exhibiting grocery in such a visually appealing approach may produce a more satisfying attitude towards the products. Wells et al. (2011) find that visual appeal can influence product quality and consumer's purchase intentions. Visual appeal reflects consumers' visual attraction by a product through grocery photo, which delivers customers with relevant product information and enables them to evaluate the performance and quality of grocery and thus form the positive attitude towards the product. Most previous studies show that visual appeal can indirectly influence consumers' impulse purchase, mediated by other variables (Wells et al., 2011), such as attitude towards product. Exhibiting grocery in such a visually appealing approach may produce more satisfied attitude towards the products. This enables us to infer that consumer easily associate grocery photo's visual appeal with the product attitudes. Therefore, I hypothesize that,

H7: Visual appeal positively affects consumers' attitude towards products.

Montoya-Weiss et al. (2003) validate that visual gravitation is positively associated with website's quality. Online visual design of the grocery store is important due to its abilities to increase website aesthetics and emotional appeal (Park et al., 2005; Liu et al., 2001; Garrett, 2003), which may in turn leads to more favorable attitude towards online store (Fiore et al., 2005). On website, ratings of visual appeal have been shown to have a strong influence on user satisfaction (Lindgaard and Dudek, 2003), and usability (Jennings, 2000). Visual appeal relates to the demonstration of visual design elements such as artifacts or graphics, which can enhance the overall presentation of a Website (Parboteeah et al., 2009). Therefore, in this study, I hypothesize that,

H8: Visual appeal positively affects consumers' attitude towards online grocery store.

Online consumers' affective reactions are provoked by some specific stimuli that are exposed to them. Furthermore, the lack of direct physical touch makes consumers more likely and directly influenced by photo decoration cues. Affective reaction gathered from photo decoration cues, in other words, to some extent leads to enjoyable experience, which potentially determines consumers' subsequent behaviors (Kim and Moon, 2009). Previous studies verify that consumers' shopping behavior has been significantly influenced by their perceived shopping enjoyment in the physical shopping experiences (Babin et al., 1994), especially in the environment of product trial, as consumer can directly contact physical goods (Kempf, 1999). Similarly, research in online context has also confirmed that consumers' shopping enjoyment is a pivotal link accompanying their online shopping experiences (Koufaris, 2002; Jarvenpaa and Todd, 1996 1997). With greater enjoyment, consumers will participate more actively in information process (Griffith et al., 2001; Andrews and Shimp, 1990), which leads to a greater possibility for high quality product.

H9: Shopping enjoyment positively affects consumers' attitude towards products.

Most prior studies have established that perceived enjoyment could positively influence consumers' attitude towards website (Lee et al., 2005; Childers et al., 2001; Monsuwe et al., 2004). Evidences show that the enjoyment experienced from e-commerce environments has a direct effect on the consequent responses in online shopping (Fiore et al., 2005a; Eroglu et al., 2003; Menon and Kahn, 2002). Menon and Kahn (2002) find that consumers who experience higher levels of pleasure from the website display higher levels of approach responses towards the website, including revisit tendency.

Raney et al. (2003) have suggested that entertainment features are helpful in promoting consumers' shopping enjoyment and improving their attitudes towards purchasing at a website, and thus increase their intention to revisit the website. Subsequently, researches have certified that enjoyment can clearly impact consumers' attitudes towards online vendors and their website (Lee et al., 2005; Childers et al., 2001; Van der Heijden, 2003; Monsuwe et al., 2004). Perceived enjoyment can positively influence user attitudes towards a system interface (Griffith et al., 2001; Jiang and Benbasat, 2007b; Morrison and Vogel, 1998) and lead to higher level of customers' online loyalty (Cyr et al., 2009), more behavioral intention to adopt a system (Van der Heijden, 2004) and greater likelihood of revisiting a website (Kourfaris, 2002). Jiang and Benbasat (2007b) also propose that shopping enjoyment positively influence consumers' attitude towards website.

As a consequence, the more perceived enjoyment from the shopping experience, the more likely that consumer would like to do shopping in those online stores. Therefore,

H10: Shopping enjoyment positively affects consumers' attitude towards online grocery store.

3.1.4 Consumers' Purchase Intention

Most research work associated with consumers' purchase behavior has supposed that purchase intention is the direct outcome of consumers' attitudinal assessment towards product. The attitude towards product refers to consumers' holistic assessments about product and the likelihood intention to complete purchases on a particular website (Jiang and Benbasat, 2007b).

It is assumed that consumers' purchase intentions are determined by two essential elements: the attitude towards product and attitude towards store. Usually, positive attitudes towards a product result in higher intention to purchase just because consumers realize the benefits of consuming the product (Fishbein and Ajzen, 1975). Jiang and Benbasat (2007b) suggest that consumers' attitude towards product has a positive impact on their intention to purchase from a website. A vivid grocery product presentation with decoration cues potentially evokes consumers' interest and intense impulse which may be changed into increased purchase intention. Such that, in this research context, it is hypothesized that,

H11: Consumers' attitudes towards product positively affect their purchase intention from a grocery store.

Similarly, Jiang and Benbasat (2007b) announce the concept of consumers' attitude towards website, that is, consumers' general evaluations of a shopping experience towards a specific website. Positive attitude towards shopping has the potential to elicit consumers to perform shopping behavior, and to enhance the possibility for purchase (Heijden et al. 2003; Jarvenpaa et al., 2000). Jiang and Benbasat (2007b) verify that consumers' attitudes towards shopping at a website significantly affect their intention to purchase a product. Therefore, I hypothesize that,

H12: Consumers' attitudes towards grocery store positively influence their purchase intention.

3.2 Research Model

Based on the Media Richness Theory (Daft and Lengel, 1986), visual design literature, the Theory of Visual Rhetoric (Scott, 1994) and literature review about photo decoration cues, a research model illustrated in Figure 3.1 is proposed. The research model aims to show how grocery photo decoration cues induce consumers' affective perception, attitudes and behavioral intention.

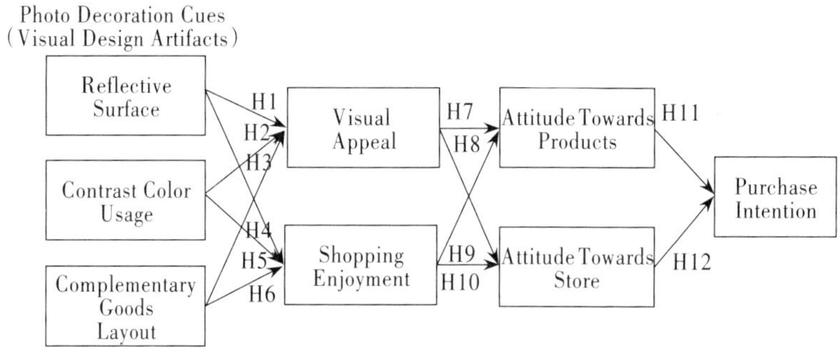

Control variables: Information deception detection, Issue involvement, Issue expertise. Privacy concern (CMV, Marker).

Figure 3.1 Research Model

The research model specifies the relationship between photo decoration cues (reflective surface, contrast color usage and complementary goods layout) and consumers' affective responses (visual appeal and shopping enjoyment), and the antecedents of purchase intention in this model are attitude towards product and attitude towards store. To account for other influences on the core dependent and mediating variables, three control variables, such as information deception detection, issue involvement, and issue expertise must be controlled. The privacy concern should be measured as the marker variables to access the common method variance (Son and Kim, 2008).

This study is guided by the following questions:

Q1. What are the photo decoration cues in online grocery?

Q2. Could photo decoration cues significantly affect the consumers' affective perception (i.e. visual appeal and shopping enjoyment), attitude (i.e. attitude towards product and store) and behavioral intention (i.e. purchase intention) of consumers?

The hypotheses of this study are summarized as follows. See Table 3.1.

Table 3.1 Hypotheses

No.	Hypotheses
H1	Reflective surface positively affects online consumers' visual appeal.
H2	Contrast color usage positively affects online consumers' visual appeal.
H3	Complementary goods layout positively affects online consumers' visual appeal.
H4	Reflective surface positively affects online consumers' shopping enjoyment.
H5	Contrast color usage positively affects online consumers' shopping enjoyment.
H6	Complementary goods layout positively affects online consumers' shopping enjoyment.
H7	Visual appeal positively affects consumers' attitudes towards products.
H8	Visual appeal positively affects consumers' attitudes towards online grocery store.
H9	Shopping enjoyment positively affects consumers' attitudes towards products.
H10	Shopping enjoyment positively affects consumers' attitudes towards online grocery store.
H11	Consumers' attitudes towards product positively affect their intention to purchase from a grocery store.
H12	Consumers' attitudes towards grocery store positively affect their purchase intention.

Chapter 4 Research Methodology

A laboratory experiment is adopted to test the effects of photo decoration cues on consumers' affective responses, attitudes and behavior intentions. In the experiment, the independent variables, mediate variables, dependent variables and possibly some other confounding variables are closely controlled to ensure the higher internal validity (Singleton and Straits, 1999). In order to increase the universality of experiment (Singleton and Straits, 1999) and the generalizability of the findings, a piece of beefsteak is selected as the representative grocery product to present. Because beefsteak is a kind of popular grocery product that common consumers would be familiar with and to which photo decoration cues could be applied easily and naturally.

4.1 Experiment Design

4.1.1 Laboratory Experiment

This study employs a laboratory experiment to validate the proposed research model and test hypotheses with a 2×2×2 factorial design. Laboratory experiment is a kind of test method through which some rigorous conditions are controlled to validate a known truth or to test the validity of hypothesis. Compared with such methods as quasi-experiment and survey methods, laboratory experiment could easily put the exogenous variables under control, which is beneficial for the internal validity (Campbell and Stanley, 1963). Laboratory experiment method is believed to be the most appropriate approach which can be adopted to simulate different control conditions and thus to observe the corresponding results (Keppel, 1991). According to Kirk (1995), the laboratory experiment allows adopting experimental elements with different levels and along with the related statistical analysis. A laboratory experiment design includes a lot of inter-related activities.

First, it should provide serial scientific hypotheses which must concern some statements about parameters or the functional form of a population. Second, the different treatment levels of independent variables can be manipulated and dependent variables' measurement must be recorded. At the same time, necessary measures must be taken to control the extraneous variables. Third, it is necessary to offer detailed introduction about the experiment factors and the related population from which data can be sampled. Fourth, the experiment must permit the random assignment of the experiment factors to different treatment levels. Last, suitable statistical analysis methods must be applied after the above process.

Given all those, a laboratory experiment design must verify the independent variables, dependent variables and exogenous variables in order to guarantee the random assignment and successful statistical methods.

The main idea of an experiment design is randomization which holds three aims. The first aim is to distribute participants' characteristics to different treatment levels to ensure there is no selective bias in the experiment results. The second aim is to allow an unbiased estimate of error to control error effect. The research designer can generate two or more similar groups of participants by random assignment. In summary, the design of the experiment identifies independent, dependent variables, disturbs variables, and indicates the way in which randomization and statistical aspects of an experiment are to be carried out. In this laboratory experiment, the three independent variables are reflective surface, contrast color usage, and complementary goods layout. Visual appeal, shopping enjoyment, attitude towards product and attitude towards store are studied and measured as mediating variables. Purchase intention is measured as dependent variable.

Two experiments are adopted in this study. The first experiment aims to confirm that three photo decorative cues are confirmed and accepted by consumers, which guarantees that this manipulation would also be effective and successful in the main experiment. In the main experiment, reflective surface, contrast color usage and complementary goods layout at two levels (i. e. presence or none) are manipulated respectively. On the post experimental questionnaire survey, participants are asked to evaluate the extent to which each item is experientially validated by photo decoration cues on a seven-point Likert scale. A virtual grocery store is developed with eight webpages and each webpage presents the same piece of beefsteak followed by the different treatments towards three photo decoration cues. The experiment is conducted completely online and participants can perform the task from any computer or mobile phone which can be connected to the Internet. After completing the shopping task (selecting a piece of beefsteak), participants are required to complete a questionnaire survey about their shopping experiences.

4.1.2 Stimuli

To perform the laboratory experiment, a virtual online grocery store is made with eight different webpages of the same product (a piece of beefsteak). Each webpage includes a photo with detailed information about the beefsteak (such as brand, source, features, price and weight) for the convenience of the participants. Each cue is manipulated with two different levels: none and presence, which means the cue exists or does not exist. There are eight different ways of combination and each combination represents one treatment. Appendix A provides the examples of the webpage for the beefsteak. Participants can use any kind of browser to browse the website. A piece of beefsteak is chosen as a grocery representative and participants could examine decoration cues, such as reflective surface, contrast color usage and complementary goods layout. They could directly examine the decoration cues associated with the beefsteak's quality per se, without regard to other information such as price, brand and source that might affect their perceptions of product quality.

The experimental steak is kept consistently across different treatments. The only difference among the treatments is the variance of visual decoration cues, and see Figure 4.1. Recognizing that the size of the product photos might influence consumers' reactions during online shopping (Lee and Benbasat, 2003; Reeves and Nass, 1996), in order to avoid this potential negative effect, the photos on the webpage are presented in the same size (5.0 cm ×6.5 cm). The aim of grocery store experiment design is to provide a shopping atmosphere that could inspire participants to imagine that they are facing an online purchase environment. The experiment task is to choose a piece of beefsteak that participants would consider buying.

The photos with different levels of reflective surface, contrast color usage and complementary goods layout are adopted. Thus, the research model is tested in a questionnaire survey including a piece of beefsteak photo

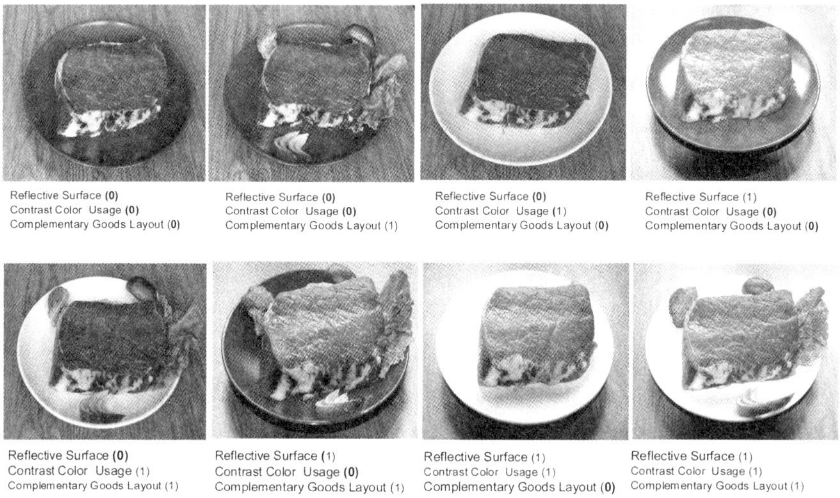

Note: 1_Presence 0_None

Figure 4.1 Different Treatments of Beefsteak Associated with the Experimental Conditions

with eight different kinds of decoration cues combination conditions: (1) Beefsteak without reflective surface, contrast color usage and complementary goods layout. (2) Beefsteak with reflective surface, without contrast color usage and complementary goods layout. (3) Beefsteak with contrast color usage, without reflective surface and complementary goods layout. (4) Beefsteak with complementary goods layout, without reflective surface and contrast color usage. (5) Beefsteak with reflective surface and contrast color usage without complementary goods layout. (6) Beefsteak with reflective surface and complementary goods layout without contrast color usage. (7) Beefsteak with contrast color usage and complementary goods layout without reflective surface. (8) Beefsteak with reflective surface, contrast color usage and complementary goods layout.

4.2 Measurement Items

4.2.1 Instrument Design

The instrument design is adopted to collect data in two parts. The first part is the information for all variables and the second part is consumers' demographic information. In the first part, scales for each variable are adopted from prior researches and adapted for the context of this study and all the item scales are positively worded. This is because some research indicates that the using negative description in developing measurement items is difficult to understand, especially in Likert scales condition (Jayawardhena, 2004). Moreover, studies suggest that negative descriptions are also believed to cause ambiguity and consequently result in some doubt towards the instruments' validity (Banerjee, 2002).

A lot of researchers (Lewis and Mitchel, 1990; Babakus and Boller, 1992) recognize these consequences and recommend the adoption of positively worded items. Based on the above consideration, all measurement items in this dissertation are positively worded.

In order to develop the measurement items, the related prior literatures are reviewed. Although many prior measurements for variables are available, some statements are not adopted directly: some believe improper items must be adapted or modified to suit the research context, and some new items should be added to satisfy the context of online grocery shopping. Hair et al. (2010) suggest that it should be equipped with sufficient number of items to capture the construct domain. Devellis (2003) also emphasizes the importance of sufficient numbers of items. If there are some overlapping items or seemingly irrelevant items, they must be erased in the final scale. In line with a rule of thumb advocated by Hair et al. (2010), a construct can be reflected at least by three items. Therefore, in this study, measurement items are adopted from many relevant sources to improve constructs' reliability and validity. Each construct's item adapted from the literature resources is shown in Table 4.1.

Table 4.1 Sources for Measurement Item Scales Development

Model Construct	Source
Visual Appeal	Loiacono et al.（2007） Mathwick et al.（2001）
Shopping Enjoyment	Verhoef and Langerak（2001） Hansen（2006） Koufaris（2001—2002） Ramus and Nielsen（2005）
Attitude Towards Product	Jiang and Benbasat（2007b）
Attitude Towards Store	Jiang and Benbasat（2007b） Grazioli and Jarvenpaa（2000）
Attitude Towards Online Grocery Shopping	Hansen et al.（2004） Hansen（2006）
Purchase Intention	Coyle and Thorson（2001） Jiang and Benbasat（2007b）
Privacy Concern	Son and Kim（2008）
Issue Involvement	Sussman and Siegal（2003）
Issue Expertise	Sussman and Siegal（2003）
Information Deception Detection	Xiao（2010）

The instruments used to collect data consist of two sections: general information and information for all variables. First, participants are asked to evaluate three photo decoration cues (reflective surface, contrast color usage and complementary goods layout) on consumers' affective reactions and totally 9 questions are used. Visual appeal and shopping enjoyment are measured using 5 items and 4 items respectively. Attitude towards product and attitude towards store are measured using 3 items and 4 items. Purchase intention is measured using 5 items. Privacy concern is measured using 4 items and issue involvement is measured using 2 items. Issue expertise is measured using 2 items and information deception detection is measured using 3 items. Then, participants are asked to answer the demographic information. At last, 331 usable responses are retained and valid return rate is 91.2%.

4.2.2 Dependent Variables

Questionnaire is carried out after each participant finishes the assigned task based on their experiment impression. Measurement items for visual appeal are adapted both from Loiacono et al. (2007) and Mathwick et al. (2001). Items for shopping enjoyment are drawn from Koufaris (2001—2002) to fit the online context. Attitudes towards store are measured by adopting established items (Coyle and Thorson, 2001; Grazioli and Jarvenpaa, 2000). Items for attitude towards product are drawn from Jiang and Benbasat (2007b). Items for purchase intention are drawn from Coyle and Thorson (2001), Gefen and Devine (2001) and Jiang and Benbasat (2007b).

4.2.3 Control Variables

To exclude other impact on the core dependent and mediating variables, a robust control is adopted to the research model. Such variables as information deception detection (Lee, 2015), issue involvement and issue expertise (Sussman and Siegal, 2003) must be controlled. These variables are regarded as the exogenous variables which can influence the information adoption process of information systems users (Sussman and Siegal, 2003). The privacy concern is also measured as the marker variable to access the common method variance (Son and Kim, 2008).

4.2.4 Likert Scale

The Likert scale is adopted to evaluate each measurement in a survey. In this study, the Likert scale adopted is the format of 7 response levels despite the fact that previous literatures employed 5 point Likert scales. There are two reasons for adopting 7 point Likert scales. Firstly, the 7 point response provides more flexibility to evaluate participants' responses level towards measurement items than 5 point scale (Tharenou et al., 2007). Namely, participants can subtly evaluate each response in a larger wide range of scales (Parasuraman, 1991; Malhotra, 1996). In addition, larger wide range of scale is helpful to generate more variances and consequently lead to increased reliability (DeVellis, 1991; Nunnally, 1978). Secondly, Malhotra (1996) argues that, more categories like scale 7 or more point scale are appropriate when the complex statistical methods are employed to analyze data. Meanwhile, correlation coefficients are influenced with the variance of number of scale points. Generally speaking, the less number of scale points, the lower correlation coefficient will be caused.

Consequently, in this study, all survey items are measured with 7 point Likert scales except the measurements of demographic information. Thus, the participants are asked to choose one of the seven responses from extremely disagree to extremely agree. Extremely disagree represents the least score (1 point) and extremely agree represents the highest score (7 point).

4.2.5 Questionnaire Design

The purpose of survey is listed on the first page of questionnaire which is shown clearly to participants. Besides, the introduction of this study and some other requirement are also presented. The questionnaire guarantees participants' privacy concern and thanks participants for participating. The main questionnaire consisted of two parts: consumers' opinions about online grocery product presentations and consumers' demographic characteristics. The questionnaire follows the accepted agreement for questionnaire design because it is a quicker and more convenient approach to seek for opinions than to ask them to give their advice. It also provides participants an opportunity to feel acceptable and comfortable with the questionnaire before involving with more complex issues.

4.3 Experiment Procedure

In this study, the whole experiment procedure consists of three sub-procedures: pretest one, pretest two and the main experiment. Pretest one mainly focuses on optimizing and refining survey items to guarantee the clarity and completeness of the instrument. The aim of pretest two is to confirm three photo decoration cues are acceptable and discernable by consumer. The main experiment aims to develop experiment website with the true stimulus.

4.3.1 Pretest One

Before the formal survey applied, a pretest is conducted with the initial version of the instruments. The main objective of the pretest one is to optimize survey items, as well as to determine whether certain questions could be loaded to the specific variables. Four professors on information systems and three experts at marketing review the instruments for face validity. The original version of the survey instrument is pretested by six professors from a Business University, each holding rich expertise in the electronic commerce research area. After gathering the feedback from those experts, the measurement items of visual appeal, attitude towards store are adapted. A focus group of twelve people also evaluates the instrument and offers good suggestion involving questions' accuracy, questions' fitness and the length of the instrument. Also, the control variables such as issue involvement and issue expertise are adapted. In addition, the participants' opinions, with regard to the manipulation of three independent variables are considered, thus guarantee the stimulus material could be solidified before the real experiment takes place.

The pretest results are reviewed by the researcher for clarity and completeness. Modifications are also made to refine and shorten the instrument. The total of 5 items for visual appeal are adapted and adopted from Loiacono et al. (2007) and Mathwick et al. (2004); 4 items measuring construct of shopping enjoyment come from Koufaris (2001—2002) with slight modifications in order to fit the research context; 3 items for attitude towards product are adopted from Jiang and Benbasat (2007b); 4 items designed to capture the attitude towards store are modified based on Grazioli and Jarvenpaa (2000), Coyle and Thorson (2001) respectively; 5 items of purchase intention adapted for its suitability to this dissertation are from Coyle and Thorson (2001), Jiang and Benbasat (2007b). All measurement items are measured on a scale 7 Likert, from "strongly disagree" to "strongly agree". The measurement items are shown in Table 4.2. The final items for the questionnaire in Chinese see Appendix C.

Table 4.2 Items for Measuring Constructs

Constructs	Measurement items	Sources
Visual Appeal (VA)	(VA1) The online grocery photo is visually pleasing. (VA2) The online grocery photo displays a visually pleasing design. (VA3) The online grocery photo is aesthetically appealing. (VA4) The way the grocery photo displays its products is attractive. (VA5) I like the way the online grocery's photo looks.	Loiacono et al. (2007) Mathwick et al. (2001)
Shopping Enjoyment (SE)	(SE1) I find my experience with this website interesting. (SE2) I find my experience with this website enjoyable. (SE3) I find my experience with this website exciting. (SE4) I find my experience with this website fun.	Koufaris (2001—2002)

续表

Constructs	Measurement items	Sources
Attitude Towards Product (ATP)	(ATP1) The product that I' ve just examined is good. (ATP2) I have formed a favorable impression towards the product that I' ve just examined. (ATP3) I like the product that I' ve just examined.	Jiang and Benbasat (2007b)
Attitude Towards Store (ATS)	(ATS1) I like shopping on this online grocery store. (ATS2) Shopping on this online grocery store is a good idea. (ATS3) Shopping on this online grocery store is appealing. (ATS4) I think this grocery store is favorable.	Jiang and Benbasat (2007b) Grazioli and Jarvenpaa (2000)
Purchase Intention (PI)	(PI1) It is likely that I will buy this grocery item. (PI2) I will definitely try this grocery item. (PI3) I have strong intentions to purchase the grocery item. (PI4) The probability that I would consider buying this grocery from website is high. (PI5) If I were to buy grocery item, I would consider buying it from this store.	Coyle and Thorson (2001) Jiang and Benbasat (2007b)
Privacy Concern (PC)	(PC1) I am concerned that the information I submit on the Internet could be misused. (PC2) I am concerned that a person can find private information about me on the Internet. (PC3) I am concerned about providing personal information on the Internet, because of what others might do with it. (PC4) I am concerned about providing personal information on the Internet, because it could be used in a way I did not foresee.	Son and Kim (2008)

续表

Constructs	Measurement items	Sources
Issue Involvement (IV)	(IV1) I am much involved in the topic of photo decoration cues in online grocery store. (IV2) Many of the issues discussed about photo decoration cues would be on my mind lately.	Sussman and Siegal (2003)
Issue Expertise (EP)	(EP1) I am much informed on the subject matter of the photo decoration cues in online grocery store. (EP2) I am an expert on the topic of photo decoration cues in online grocery store.	Sussman and Siegal (2003)
Information Deception Detection (IDD)	Overall, I believe that the given product presented in this online grocery store are… (IDD1) Misleading. (IDD2) Deceptive. (IDD3) Biased.	Xiao (2010)

Note: 'VA' stands for visual appeal. 'SE' stands for shopping enjoyment. 'ATP' stands for attitude towards product. 'ATS' stands for attitude towards store. 'PI' stands for purchase intention. 'PC' stands for privacy concern. 'IV' stands for issue involvement. 'EP' stands for issue expertise. 'IDD' stands for information deception detection.

4.3.2 Pretest Two

The second pretest is designed to confirm that three photo decoration cues are acceptable and clearly discernable by consumers. Various techniques, such as camera control, focusing, flicker, exposure and balancing are adopted to polish and render the beefsteak. In other words, reflective surface, contrast color usage and complementary goods layout of grocery photo can be clearly confirmed by consumers, which are to guarantee the main experiment will be successful. In pretest two, the experiment is divided into three different contrast conditions. In the first condition, two contrast grocery photos with the same beefsteak are needed to provide. A special photo technology is adopted to polish the beefsteak's reflective surface effect by changing the camera' s brightness and flicker towards the surface of beefsteak, which can better lead to reflective surface effect in the first photo. On the contrary, there is no reflective surface effect on the beefsteak in the second photo. Then, in the second condition there are also two photos. Onion, mushroom, cauliflower and garlic are placed as the background goods of the beefsteak to construct the complementary goods layout effect in the first photo, there is no complementary goods in the second photo as the contrast photo. In the third condition, a red plate is used as background of beefsteak then a photo is taken. The red plate is replaced with white plate to take another photo as the contrast photo, which aims to make contrast color usage effect in grocery photo.

For a manipulation check, this experiment is conducted with 148 participants who are demographically similar to the main experiment participants. Participants are divided into three groups and asked to rate whether there are some effects of reflective surface, contrast color usage and complementary goods layout in each condition.

In the experiment, one group is assigned to the experiment with contrast effect, while another group responses to no contrast effect treatment. Three pairs of group are asked to verify whether the given beefsteak holds the decorative effects or not. After each distinguished photo group, the participants completed questionnaires, responses are recorded on a "Yes" or "No". The questionnaire and screenshots of the manipulation for the experiment are found in the Appendix D. The results validated the success of the manipulation in laboratory experiment.

In the first control group experiment, for the beefsteak with lighting and shining effect, 90.8% of the participants from experiment group agree that there is reflective surface effect on the surface of beefsteak. In other words, only 9.2% participants believe there is no such effect. On the contrary,

87.5% of the participants from contrast group argue that there is no reflective surface effect on the beefsteak without lighting and shining. In the second group, 78.9% of the participants from experiment group view the beefsteak with a white plate as the effect of contrast color usage, which means 21.1% participants don' t think so. While 86.1% of the participants from contrast group consider that there is no contrast color between beefsteak and red plate. Similar to the third group condition, 82% of the participants from experiment group view beefsteak with onion, mushroom and garlic as the complementary goods layout effect (18% of the opponents), while the contrast group's rating of no complementary effect score reaches 100% in the condition of beefsteak without any background goods. The experiment manipulation check results are shown in Table 4.3.

Table 4.3 Manipulation Check Results

	Reflective Surface	Contrast Color Usage	Complementary Goods Layout
Experimental Group	9.2%	21.1%	18%
Control Group	87.5%	86.1%	100%

All above results are consistent with the expected effects of reflective surface, color usage and complementary goods layout thereby ensuring the manipulation is successful. Hence, identification of reflective surface, contrast color usage and complementary goods layout in the grocery photo are valid and successful.

4.3.3 Main Experiment

The selection of the sample is an important decision, because the size of the sample and sample selection are two critical indicators and they should be suitable for research purpose and statistic methodology adopted in the study. For example, using smaller sample will lead to more convergent problems that decrease the precision of parameter estimates (Loehlin, 1992).

There are some guidelines on how to determine the appropriate sample size in the past studies (Fabrigar et al., 1999; Gorsuch, 1983). These guidelines focus on the number of measured variables, in other words, the more measured variables the larger sample size is needed. Gorsuch (1983) recommends a ratio of 5 participants each variable and the sample size should be larger than 100, but Nunnally (1978) suggests the ratio of 10 to 1 is better. According to Hair et al. (1998), a sample size from 100 to 200 is popular. While Nunnaly (1978) proposes that a sample of 300 participants is enough to test research model. Based on the Nunnaly's recommendation, the sample size for this study is 331.

The research object of current study is to explore grocery photo decoration cues on customers' affective response in online grocery store. The target population of this study is Chinese consumers who have shopped online grocery in the last few months. It is impractical to collect a complete list of consumers across China who has had an experience of online grocery shopping. A Business University in China is chosen where students and staff voluntarily participated in the experiment. Hence, they are deemed to be suitable participants for this study. University students and staff who have experienced grocery shopping are recruited to participate in this study.

Some researchers question the appropriateness of using students as participants (Gordon et al., 1986), others (Greenberg, 1987; McKnight et al., 2002; Legris et al., 2003) suggest that students participants used in this type of research would not affect the validity and generalizability of findings. Therefore, a total number of 331 University students and staff are recruited as participants in this study. Students and staff are randomly assigned to one of the eight different experiment groups. At the end of the experiment, each student or staff is offered $8 to encourage their participation in the experiment.

At first, participants are given a brief introduction about the purpose of the survey in the University computer labs where they have easy access to the

Internet. Participants are notified that (1) the study is just for academic purpose, and all of their feedback will remain anonymous; (2) It emphasizes that there are no right or wrong answers, so they are asked to answer the questions honestly and impersonally, because only their opinions are needed; (3) Imagine there is a virtual online grocery store (a website) where you will do your grocery shopping on this website; (4) The security of participants' personal information is guaranteed; (5) This survey may be finished in 15~20 minutes.

Next, the students and staff are randomly assigned to one of the following experiment conditions: (1) Beefsteak without reflective surface, contrast color usage and complementary goods layout. (2) Beefsteak with reflective surface, without contrast color usage and complementary goods layout. (3) Beefsteak with contrast color usage, without reflective surface and complementary goods layout. (4) Beefsteak with complementary goods layout, without reflective surface and contrast color usage. (5) Beefsteak with reflective surface and contrast color usage without complementary goods layout. (6) Beefsteak with reflective surface and complementary goods layout without contrast color usage. (7) Beefsteak with contrast color usage and complementary goods layout without reflective surface. (8) Beefsteak with reflective surface, contrast color usage and complementary goods layout. The different purchasing scenario is presented to the participants in each group, where participants are assumed to purchase a piece of beefsteak from one experiment webpage. In the experiment, participants are exposed to a piece of beefsteak photo with different decoration cues which enable them to evaluate the grocery respectively.

After that, participants in each group are given the URL of questionnaire to do online survey and are asked to inspect the beefsteak as if they were shopping and had to decide whether to buy or not. At the same time, the participants are told that they would get $8 as a reward for their participation and be required to be consulted after collecting the data.

After the participants complete the questionnaires, responses are recorded on a seven-point scale with "extreme disagree" to "extreme agree" and the structured questionnaire survey can be seen in Appendix E.

Chapter 5 Data Analysis and Results

This study employs three photo decoration cues with different levels of treatment to validate consumers' affective perception, attitudes and behavior. To examine the research model and hypotheses, the research model is divided into two parts and thus data analysis is conducted in two steps. In the first step, I challenge to perform ANOVA to check the impacts of photo decoration cues on consumers' affective perceptions. In the second step, I attempt to perform the structural equation model analysis using PLS to validate the consumers' effective perception towards attitudes and purchase intention. Why is 2-step analysis method adopted in this study? This is because that the independent variables could not be directly measured, thus, this study makes the instruments to reflect each independent variable's characteristics and assigns the binary type data to them, which are different from the Likert data of the dependent variables. Consequently, this study challenges to perform two steps approach to analyze the research model and validate hypotheses. Data analysis is conducted with SPSS 21.0 software and the structural equation modeling is tested by smartPLS 2.0 software.

5.1 Primary Analysis

Preliminary analysis, such as the demographics of the participants and general statistics must be firstly processed. The descriptive analysis generally includes descriptive statistics information such as standard deviation, mean, and sum.

5.1.1 Demographics of Participants

Among the 331 participants, 52.6% are female and 47.4% are male, which shows that women are more likely to shop grocery product than men. This partly reflects that women are main shoppers in the family (Mintel, 2011) .Participants mainly are in the age group of 20~35 years (92.4%) and 36~45 years (3.6%) .

The findings show that the major online grocery shoppers are between the age of 20 and 45 years. In other words, the younger shoppers are controlling the market while the older shoppers are the least likely to purchase grocery product. The result is similar with the findings of Mintel (2011) which argues that the age between 25 and 44 is the most likely to purchase online. Shopping frequency is divided into four groups, the largest shopping frequency group is those who shops a few times a year with 171 participants (51.5% per year), followed by a few times a month with 62 (18.8% a month), a few times of week with 40 (12.1% a week) and a few times half a year with 38 (11.5% half a year) . The demographic information of the participants is shown in Table 5.1.

Table 5.1 Demographics of Participants

Characteristics		Statistics （%）
Gender	Male	157 （47.4%）
	Female	174 （52.6%）
Age	20~35 years	306 （92.4%）
	36~45 years	12 （3.6%）
	46~55 years	8 （2.4%）
	56 plus years	5 （1.5%）
Shopping Frequency	A few times a week	40 （12.1%）
	A few times a month	62 （18.8%）
	A few times half a year	38 （11.5%）
	A few times a year	171 （51.5%）

5.1.2 Descriptive Statistics

The descriptive statistics information for all the variables generally contains number of participants, standard deviation and the mean. Table 5.2a shows the means and standard deviations for all variables.

Table 5.2a General Descriptive Statistics

Construct	N	Mean	Std. Deviation
RS	331	0.5106	0.50065
CCU	331	0.4924	0.50070
CGL	331	0.5015	0.50075
VA	331	4.4798	1.50003
SE	331	4.3119	1.46057
ATP	331	4.6425	1.58531
ATS	331	4.3648	1.46462
PIN	331	4.0689	1.52406

Note: ‘RS’ stands for reflective surface. ‘CCU’ stands for contrast color usage. ‘CGL’ stands for complementary goods layout. ‘VA’ stands for visual appeal. ‘SE’ stands for shopping enjoyment. ‘ATP’ stands for attitude towards product. ‘ATS’ stands for attitude towards store. ‘PIN’ stands for purchase intention.

The construct of reflective surface, contrast color usage and complementary goods layout are the most important variables that influence consumers' affective responses about grocery product and online grocery store. Consequently, they have higher means of 0.5106, 0.4924 and 0.5015 compare with other variables. Table 5.2b reveals the descriptive information based on three photo decorations cues' different levels.

Table 5.2b Descriptive Statistics Based on Three Cues' Different Levels

Measurement		Reflective Surface		Contrast Color Usage		Complementary Goods Layout	
		Yes	No	Yes	No	Yes	No
Visual Appeal	Mean	4.99	3.95	5.03	3.95	5.29	3.67
	(S.D)	1.52	1.29	1.26	1.53	1.16	1.36
Shopping Enjoyment	Mean	5.44	3.14	4.37	4.25	4.83	3.79
	(S.D)	0.93	0.88	1.42	1.49	1.48	1.25

5.1.3 Control and Manipulation Checks

A control check is done to guarantee that there are no significant differences in participants' age, gender, shopping frequency across the different treatment conditions. Based on the measurement items' evaluation of visual appeal and shopping enjoyment, data are analyzed to test whether participants understand the experiment requirements and whether three decoration cues can stimulate consumers' affective responses. Those participants who fail to complete the experiment task are removed from the statistical analyses, which thus results 331 valid data sets which are far beyond the requirement of partial least squares (PLS) evaluation standard.

Among the 331 participants, 52.6% are female and 47.4% are male. The independent t-test is verified that there is no significant difference in gender towards visual appeal ($t=-0.387$, $MD=-0.06395$, $p=0.699$) and shopping enjoyment ($t=1.794$, $MD=0.28745$, $p=0.074$), results are shown in Table 5.3. ANOVA reveals that there is no significant difference between the groups in terms of participants' age ($F=1.661$, $p=0.175$; $F=0.894$, $p=0.445$) and the frequency of shopping grocery ($F=0.371$, $p=$

0.774; F=0.385, p=0.764), results are displayed in Table 5.4 and Table 5.5. The correlation coefficients between control variables and visual appeal and shopping enjoyment are not significant, and see Table 5.6. Therefore none of the control variables has a significant impact on the dependent variables in this study. So, manipulation check verified that the randomly assigned participants to the different experimental groups is successful.

Table 5.3 Independent Sample Test for Gender

		Levene's Test for Equality of Variances		t-test for Equality of Means						
		F	Sig.	t	df	Sig. (2-tailed)	Mean Difference	Std. Error Difference	95% Confidence Interval of the Difference	
									Lower	Upper
VA	Equal variances assumed	0.480	0.489	-0.387	329	0.699	-0.06395	0.16533	-0.38919	0.26129
	Equal variances not assumed			-0.387	325.841	0.699	-0.06395	0.16529	-0.38912	0.26122
SE	Equal variances assumed	1.342	0.248	1.794	329	0.074	0.28745	0.16024	-0.02777	0.60266
	Equal variances not assumed			1.788	319.531	0.075	0.28745	0.16081	-0.02893	0.60382

Note: 'VA' stands for visual appeal. 'SE' stands for shopping enjoyment.

Table 5.4 ANOVA for Age

		Sum of Squares	df	Mean Square	F	Sig.
VA	Between Groups	11.143	3	3.714	1.661	0.175
	Within Groups	731.392	327	2.237		
	Total	742.534	330			
SE	Between Groups	5.724	3	1.908	0.894	0.445
	Within Groups	698.256	327	2.135		
	Total	703.980	330			

Note: 'VA' stands for visual appeal. 'SE' stands for shopping enjoyment.

Table 5.5 ANOVA for Shopping Frequency

		Sum of Squares	df	Mean Square	F	Sig.
VA	Between Groups	2.522	3	0.841	0.371	0.774
	Within Groups	740.012	327	2.263		
	Total	742.534	330			
SE	Between Groups	2.476	3	0.825	0.385	0.764
	Within Groups	701.504	327	2.145		
	Total	703.980	330			

Note: 'VA' stands for visual appeal. 'SE' stands for shopping enjoyment.

Table 5.6 Correlation Coefficients between Control Variables and Dependent Variables

Control Variables	Shopping Enjoyment	Visual Appeal
Information Deception Detection	-0.000581	-0.014491
Issue Expertise	-0.007029	-0.004473
Issue Involvement	-0.013499	-0.016180

5.2 Step One: ANOVA Test for Variation of Three Photo Decoration Cues

In the first step, ANOVA is used to test hypotheses between photo decoration cues and consumers' affective perception. The ANOVA method is employed to check and verify whether the experiment operations of different levels of reflective surface, contrast color usage and complementary goods layout are successful or not. The experiment results demonstrate that the eight different treatment conditions significantly affect photo's reflective surface ($p<0.01$), contrast color usage ($p<0.01$) and complementary goods layout ($p<0.01$). The first step of validating research model is shown in Figure 5.1.

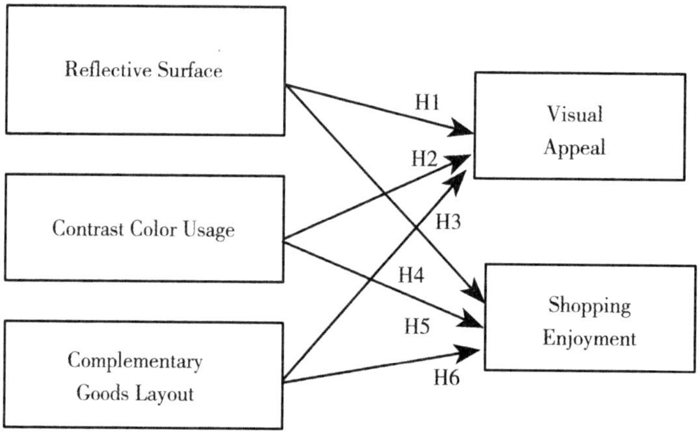

Figure 5.1 The First Step of Validating Research Model

5.2.1 Effects on Visual Appeal

The ANOVA test is conducted to verify the influence of three photo decoration cues on visual appeal and shopping enjoyment respectively. Table 5.7 and Table 5.8 list the descriptive statistics information about visual appeal and shopping enjoyment.

Table 5.7 Descriptive Statistics on Visual Appeal

RS	CCU	CGL	Group	Mean	Std. Deviation	N
No	No	No	1	2.7667	0.77007	42
		Yes	2	4.3619	1.12899	42
	Yes	No	3	3.8865	1.25215	37
		Yes	5	4.8098	0.97258	41
Yes	No	No	4	2.9864	0.90209	44
		Yes	6	5.8150	0.89028	40
	Yes	No	7	5.0810	1.04910	42
		Yes	8	6.1628	0.55421	43

Note: ‘RS’ stands for reflective surface. ‘CCU’ stands for contrast color usage. ‘CGL’ stands for complementary goods layout.

Table 5.8 Descriptive Statistics on Shopping Enjoyment

RS	CCU	CGL	Group	Mean	Std. Deviation	N
No	No	No	1	2.6964	0.79353	42
		Yes	2	3.4286	0.91613	42
	Yes	No	3	2.7365	0.71187	37
		Yes	5	3.6646	0.63396	41
Yes	No	No	4	4.8693	0.65700	44
		Yes	6	6.0688	0.64794	40
	Yes	No	7	4.6845	0.63931	42
		Yes	8	6.1570	0.63626	43

Note: ‘RS’ stands for reflective surface. ‘CCU’ stands for contrast color usage. ‘CGL’ stands for complementary goods layout.

There are eight groups representing different treatments. Each group has the following label: group 1 identifies reflective surface (0), contrast color usage (0), complementary goods layout (0); group 2 identifies reflective surface (0), contrast color usage (0), complementary goods layout (1); group 3 identifies reflective surface (0), contrast color usage (1), complementary goods layout (0); group 4 identifies reflective surface (1), contrast color usage (0), complementary goods layout (0); group 5 identifies reflective surface (0), contrast color usage (1), complementary goods layout (1); group 6 identifies reflective surface (1), contrast color usage (0), complementary goods layout (1); group 7 identifies reflective surface (1), contrast color usage (1), complementary goods layout (0); group 8 identifies reflective surface (1), contrast color usage (1), complementary goods layout (1).

For visual appeal, the ANOVA statistical analysis results shown in Table 5.9. The experiment results reveal that the p-values of reflective surface, contrast color usage and complementary goods layout are confirmed significantly ($p=0.000$) across the total eight experiment treatments. Therefore, H1 is confirmed with the construct of reflective surface which positively affects visual appeal, H2 is supported with the construct of contrast color usage which positively affects visual appeal, and H3 is also supported with the construct of complementary goods layout which positively affects visual appeal. Moreover, the results also indicate that the interaction effects among three independent variables are significant towards visual appeal. The findings show that interaction effect of complementary goods layout and reflective surface are significant (F [2, 331] =10.953, p=0.001), complementary goods layout and contrast color usage are significant (F [2, 331] =33.073, p=0.000), reflective surface and contrast color usage are significant (F [2, 331] =4.325, p=0.038), reflective surface, contrast color usage and complementary goods layout are also significant (F [3, 331] =6.531, p=0.011).

Table 5.9 Tests of Between-Participants Effects on Visual Appeal

Source	Type III Sum of Squares	df	Mean Square	F	Sig.	Partial Eta Squared
RS	91.890	1	91.890	100.687	0.000	0.238
CCU	82.962	1	82.962	90.905	0.000	0.220
CGL	213.239	1	213.239	233.653	0.000	0.420
RS * CCU	3.947	1	3.947	4.325	0.038	0.013
RS * CGL	9.996	1	9.996	10.953	0.001	0.033
CCU * CGL	30.183	1	30.183	33.073	0.000	0.093
RS * CCU * CGL	5.960	1	5.960	6.531	0.011	0.020

Note: 'RS' stands for reflective surface. 'CCU' stands for contrast color usage. 'CGL' stands for complementary goods layout. *P<0.05, **p<0.01, ***p<0.001.

Figure 5.2 graphically represents the cross relationships among the three decoration cues with different levels on visual appeal.

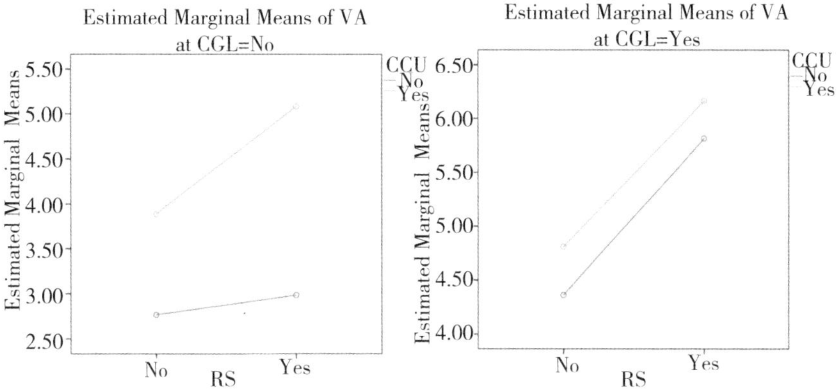

Estimated Marginal Means of VA at CGL=No　　Estimated Marginal Means of VA at CGL=Yes

Figure 5.2 Estimated Marginal Means of Visual Appeal

The ANOVA results also indicate that visual appeal has significant differences in means between group comparing based on Tukey HSD test. However, the means of group 1 and 4, group 2 and 5, group 3 and 2, group 6 and 8, group 7 and 5 are not significantly different. The Tukey HSD multiple comparison results towards visual appeal are shown in Table 5.10.

Table 5.10 Multiple Comparison Results Towards Visual Appeal

(I) Group	(J) Group	Mean Difference (I-J)	Std. Error	Sig.	95% Confidence Interval	
					Lower Bound	Upper Bound
1	2	-1.5952*	0.20847	0.000	-2.2313	-0.9592
	3	-1.1198*	0.21539	0.000	-1.7770	-0.4627
	4	-0.2197	0.20608	0.963	-0.8484	0.4091
	5	-2.0431*	0.20973	0.000	-2.6830	-1.4032
	6	-3.0483*	0.21106	0.000	-3.6923	-2.4044
	7	-2.3143*	0.20847	0.000	-2.9503	-1.6783
	8	-3.3961*	0.20725	0.000	-4.0284	-2.7638
2	1	1.5952*	0.20847	0.000	0.9592	2.2313
	3	0.4754	0.21539	0.350	-0.1817	1.1326
	4	1.3755*	0.20608	0.000	0.7468	2.0043
	5	-0.4479	0.20973	0.394	-1.0877	0.1920
	6	-1.4531*	0.21106	0.000	-2.0970	-0.8092
	7	-0.7190*	0.20847	0.015	-1.3551	-0.0830
	8	-1.8009*	0.20725	0.000	-2.4332	-1.1686

续表

(I) Group	(J) Group	Mean Difference (I-J)	Std. Error	Sig.	95% Confidence Interval	
					Lower Bound	Upper Bound
3	1	1.1198*	0.21539	0.000	0.4627	1.7770
	2	-0.4754	0.21539	0.350	-1.1326	0.1817
	4	0.9001*	0.21309	0.001	0.2500	1.5502
	5	-0.9233*	0.21662	0.001	-1.5842	-0.2624
	6	-1.9285*	0.21790	0.000	-2.5933	-1.2637
	7	-1.1945*	0.21539	0.000	-1.8516	-0.5373
	8	-2.2763*	0.21422	0.000	-2.9299	-1.6227
4	1	0.2197	0.20608	0.963	-0.4091	0.8484
	2	-1.3755*	0.20608	0.000	-2.0043	-0.7468
	3	-0.9001*	0.21309	0.001	-1.5502	-0.2500
	5	-1.8234*	0.20737	0.000	-2.4561	-1.1907
	6	-2.8286*	0.20870	0.000	-3.4654	-2.1919
	7	-2.0946*	0.20608	0.000	-2.7233	-1.4658
	8	-3.1764*	0.20486	0.000	-3.8014	-2.5514
5	1	2.0431*	0.20973	0.000	1.4032	2.6830
	2	0.4479	0.20973	0.394	-0.1920	1.0877
	3	0.9233*	0.21662	0.001	0.2624	1.5842
	4	1.8234*	0.20737	0.000	1.1907	2.4561
	6	-1.0052*	0.21231	0.000	-1.6530	-0.3575
	7	-0.2712	0.20973	0.901	-0.9111	0.3687
	8	-1.3530*	0.20853	0.000	-1.9892	-0.7168
6	1	3.0483*	0.21106	0.000	2.4044	3.6923
	2	1.4531*	0.21106	0.000	0.8092	2.0970
	3	1.9285*	0.21790	0.000	1.2637	2.5933
	4	2.8286*	0.20870	0.000	2.1919	3.4654
	5	1.0052*	0.21231	0.000	0.3575	1.6530
	7	0.7340*	0.21106	0.013	0.0901	1.3780
	8	-0.3478	0.20986	0.715	-0.9880	0.2925
7	1	2.3143*	0.20847	0.000	1.6783	2.9503
	2	0.7190*	0.20847	0.015	0.0830	1.3551
	3	1.1945*	0.21539	0.000	0.5373	1.8516
	4	2.0946*	0.20608	0.000	1.4658	2.7233
	5	0.2712	0.20973	0.901	-0.3687	0.9111
	6	-0.7340*	0.21106	0.013	-1.3780	-0.0901
	8	-1.0818*	0.20725	0.000	-1.7141	-0.4495

续表

(I) Group	(J) Group	Mean Difference (I-J)	Std. Error	Sig.	95% Confidence Interval	
					Lower Bound	Upper Bound
8	1	3.3961*	0.20725	0.000	2.7638	4.0284
	2	1.8009*	0.20725	0.000	1.1686	2.4332
	3	2.2763*	0.21422	0.000	1.6227	2.9299
	4	3.1764*	0.20486	0.000	2.5514	3.8014
	5	1.3530*	0.20853	0.000	0.7168	1.9892
	6	0.3478	0.20986	0.715	−0.2925	0.9880
	7	1.0818*	0.20725	0.000	0.4495	1.7141

Note: each group stands for the different permutation and combination of levels of three photo decoration cues. Group 1 stands for reflective surface (0), contrast color usage (0), complementary goods layout (0). Group 2 stands for reflective surface (0), contrast color usage (0), complementary goods layout (1). Group 3 stands for reflective surface (0), contrast color usage (1), complementary goods layout (0). Group 4 stands for reflective surface (1), contrast color usage (0), complementary goods layout (0). Group 5 stands for reflective surface (0), contrast color usage (1), complementary goods layout (1). Group 6 stands for reflective surface (1), contrast color usage (0), complementary goods layout (1). Group 7 stands for reflective surface (1), contrast color usage (1), complementary goods layout (0). Group 8 stands for reflective surface (1), contrast color usage (1), complementary goods layout (1).

5.2.2 Effects on Shopping Enjoyment

In terms of shopping enjoyment, the ANOVA statistical analysis results are shown in Table 5.11. The ANOVA results show that the p-value of reflective surface and complementary goods layout are significant ($p=0.000$) across the eight groups of participants, therefore, H4 is confirmed with the construct of reflective surface which positively affects consumers' shopping enjoyment. H6 is accepted as the construct of complementary goods layout which positively affects consumers' shopping enjoyment. However, the relationship between contrast color usage and shopping enjoyment is not significant (p>0.10), therefore H5 is not supported. In addition, the interaction effects of reflective surface and complementary goods layout are significant too (F [2, 331] =10.446, p=0.001), and other interaction effects are not significant (p>0.10).

Table 5.11 Tests of Between-Participants Effects on Shopping Enjoyment

Source	Type III Sum of Squares	df	Mean Square	F	Sig.	Partial Eta Squared
Corrected Model	540.738[a]	7	77.248	152.848	0.000	0.768
Intercept	6071.748	1	6071.748	12013.908	0.000	0.974
RS	441.762	1	441.762	874.096	0.000	0.730
CCU	0.166	1	0.166	0.329	0.567	0.001
CGL	96.826	1	96.826	191.586	0.000	0.372
RS * CCU	0.717	1	0.717	1.418	0.235	0.004
RS * CGL	5.280	1	5.280	10.446	0.001	0.031
CCU * CGL	1.135	1	1.135	2.246	0.135	0.007
RS * CCU * CGL	0.031	1	0.031	0.061	0.806	0.000

Note: 'RS' stands for reflective surface. 'CCU' stands for contrast color usage. 'CGL' stands for complementary goods layout. *P<0.05, **p<0.01, ***p<0.001.

Figure 5.3 graphically represents the cross relationships among the three decoration cues with different levels on shopping enjoyment.

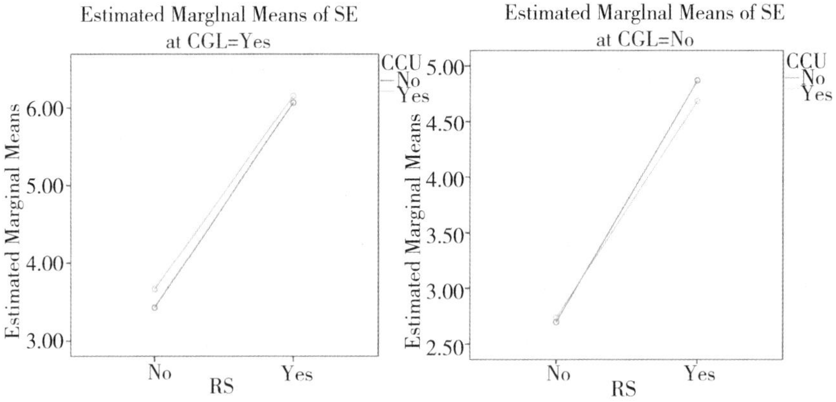

Estimated Marginal Means of SE at CGL=YES Estimated Marginal Means of SE at CGL=No

Figure 5.3 Estimated Marginal Means of Shopping Enjoyment

The ANOVA results also indicate that shopping enjoyment has significant differences in means between group comparing based on Tukey HSD test. Whereas the means of group 1 and 3, group 4 and 7, group 5 and 2, group 6 and 8 are not significantly different. The Tukey HSD multiple comparison results towards visual appeal are shown in Table 5.12.

Table 5.12 Multiple Comparison Results Towards Shopping Enjoyment

(I) Group	(J) Group	Mean Difference (I-J)	Std. Error	Sig.	95% Confidence Interval	
					Lower Bound	Upper Bound
1	2	-0.7321*	0.15513	0.000	-1.2054	-0.2588
	3	-0.0401	0.16029	1.000	-0.5291	0.4490
	4	-2.1729*	0.15336	0.000	-2.6408	-1.7050
	5	-0.9682*	0.15608	0.000	-1.4444	-0.4920
	6	-3.3723*	0.15706	0.000	-3.8515	-2.8931
	7	-1.9881*	0.15513	0.000	-2.4614	-1.5148
	8	-3.4605*	0.15423	0.000	-3.9311	-2.9900
2	1	0.7321*	0.15513	0.000	0.2588	1.2054
	3	0.6921*	0.16029	0.001	0.2031	1.1811
	4	-1.4407*	0.15336	0.000	-1.9086	-0.9729
	5	-0.2361	0.15608	0.800	-0.7122	0.2401
	6	-2.6402*	0.15706	0.000	-3.1194	-2.1610
	7	-1.2560*	0.15513	0.000	-1.7293	-0.7827
	8	-2.7284*	0.15423	0.000	-3.1989	-2.2579
3	1	0.0401	0.16029	1.000	-0.4490	0.5291
	2	-0.6921*	0.16029	0.001	-1.1811	-0.2031
	4	-2.1328*	0.15857	0.000	-2.6166	-1.6490
	5	-0.9281*	0.16120	0.000	-1.4200	-0.4363
	6	-3.3323*	0.16215	0.000	-3.8270	-2.8375
	7	-1.9480*	0.16029	0.000	-2.4371	-1.4590
	8	-3.4205*	0.15941	0.000	-3.9068	-2.9341
4	1	2.1729*	0.15336	0.000	1.7050	2.6408
	2	1.4407*	0.15336	0.000	0.9729	1.9086
	3	2.1328*	0.15857	0.000	1.6490	2.6166
	5	1.2047*	0.15431	0.000	0.7339	1.6755
	6	-1.1994*	0.15531	0.000	-1.6733	-0.7256
	7	0.1848	0.15336	0.930	-0.2831	0.6527
	8	-1.2877*	0.15245	0.000	-1.7528	-0.8226

续表

(I) Group	(J) Group	Mean Difference (I-J)	Std. Error	Sig.	95% Confidence Interval	
					Lower Bound	Upper Bound
5	1	0.9682*	0.15608	0.000	0.4920	1.4444
	2	0.2361	0.15608	0.800	-0.2401	0.7122
	3	0.9281*	0.16120	0.000	0.4363	1.4200
	4	-1.2047*	0.15431	0.000	-1.6755	-0.7339
	6	-2.4041*	0.15799	0.000	-2.8861	-1.9221
	7	-1.0199*	0.15608	0.000	-1.4961	-0.5437
	8	-2.4923*	0.15518	0.000	-2.9658	-2.0189
6	1	3.3723*	0.15706	0.000	2.8931	3.8515
	2	2.6402*	0.15706	0.000	2.1610	3.1194
	3	3.3323*	0.16215	0.000	2.8375	3.8270
	4	1.1994*	0.15531	0.000	0.7256	1.6733
	5	2.4041*	0.15799	0.000	1.9221	2.8861
	7	1.3842*	0.15706	0.000	0.9050	1.8634
	8	-0.0882	0.15617	0.999	-0.5647	0.3882
7	1	1.9881*	0.15513	0.000	1.5148	2.4614
	2	1.2560*	0.15513	0.000	0.7827	1.7293
	3	1.9480*	0.16029	0.000	1.4590	2.4371
	4	-0.1848	0.15336	0.930	-0.6527	0.2831
	5	1.0199*	0.15608	0.000	0.5437	1.4961
	6	-1.3842*	0.15706	0.000	-1.8634	-0.9050
	8	-1.4725*	0.15423	0.000	-1.9430	-1.0019
8	1	3.4605*	0.15423	0.000	2.9900	3.9311
	2	2.7284*	0.15423	0.000	2.2579	3.1989
	3	3.4205*	0.15941	0.000	2.9341	3.9068
	4	1.2877*	0.15245	0.000	0.8226	1.7528
	5	2.4923*	0.15518	0.000	2.0189	2.9658
	6	0.0882	0.15617	0.999	-0.3882	0.5647
	7	1.4725*	0.15423	0.000	1.0019	1.9430

Note: each group stands for the different permutation and combination of levels of three photo decoration cues. Group 1 stands for reflective surface (0), contrast color usage (0), complementary goods layout (0). Group 2 stands for reflective surface (0), contrast color usage (0), complementary goods layout (1). Group 3 stands for reflective surface (0), contrast color usage (1), complementary goods layout (0). Group 4 stands for reflective surface (1), contrast color usage (0), complementary goods layout (0). Group 5 stands for reflective surface (0), contrast color usage (1), complementary goods layout (1). Group 6 stands for reflective surface (1), contrast color usage (0), complementary goods layout (1). Group 7 stands for reflective surface (1), contrast color usage (1), complementary goods layout (0). Group 8 stands for reflective surface (1), contrast color usage (1), complementary goods layout (1).

Overall, based on the above discussion, the ANOVA analysis results clearly confirm that the photo decoration cues are the power source for consumers to provoke their affective reactions, followed by attitude towards product and attitude towards grocery store and then lead to their purchase intention. All hypotheses are supported by experimental analysis with the exception of contrast color usage to shopping enjoyment ($p>0.1$). As expected, reflective surface, contrast color usage and complementary goods layout of photo decoration cues significantly influence visual appeal and shopping enjoyment.

5.3 Measurement Model

5.3.1 Exploratory Factor Analysis

Before conducting the right-hand of research model test, it's necessary to pay attention to the exploratory factor analysis (EFA). EFA is widely used in extant researches which are beneficial to determine the number of factors or latent variables. It can account for correlations among many measures in a specific research model (Norris et al., 2009). EFA requires that each step should be given a strict concern in its procedure. Actually, the efficiency of EFA relies on how it is carried out (Fabrigar et al., 1999) and it doubtlessly leads to the wrong results and misleads the readers when researchers make a poor decision (Norris et al. 2009; Finch and West, 1997; Cudeck and O' Dell, 1994). Therefore, researchers should make the appropriate judgment on how to analyze the whole procedure and what results could better explain the aims of research (Norris et al., 2009) when performing the EFA. One purpose of performing the EFA is to validate a construct' s dimensionality which means that the latent features of structures can be identified by EFA, and consequently, these results could be used to evaluate construct validity. After assessing each construct of research model via EFA, researchers can continue to test the research hypotheses. Another purpose of EFA is to develop new instrument measurement scales in a survey or laboratory experiment.

Churchill (1979) states that the EFA is very important to determine the quality of the instruments. According to Chou (2011), the researcher should purify the items within each construct by examining corrected item to total correlations and communality, running EFA with interpretability criterion and checking indicator loadings. Survey items with corrected item-total correlation less than 0.3 should be dropped from the study (De Vaus, 2002). While Cho (2011) insists that items' correlation less than 0.4 should be thrown away.

There are some standards for determining the number of factors. For example, Kaiser suggests that factors with a maximum eigenvalue are larger than one (Nunnally, 1978). This study follows the interpretability criteria and the variance-explained criteria. The interpretability criterion limits the number of factors to those whose dimension of meaning is readily comprehensible. The variance-explained criterion is the percentage of total variance extracted, and how many extracted factors are accounted for within the latent variable (Gorsuch, 1983; Vogt, 2005). It is acceptable when the variance explained is around 50%~75% (Neill, 2008). Factor loading has to do with correlations between items and certain components (Gorsuch, 1983). Generally speaking, the larger loading the closer relationship among the items which can easily form a factor. De Vaus (2002) mentions that the higher the loadings, the more the variables belong to the components.

5.3.2 Confirmatory Factor Analysis

Confirmatory Factor Analysis (CFA) is a special statistical procedure most commonly employed in extant studies. CFA is adopted to validate whether a construct' s measure is in line with researchers' understanding about constructs (Kline, 2010). Therefore, the aim of CFA is to validate whether the data fit the hypothesized model which is based on theory or prior studies. According to Hair et al. (2012), in order to evaluate as to what extent that the construct is suitable to be measured by respective indicator variables, it is necessary for researchers to distinguish reflective and formative measurement approaches.

In this study, the hypothetical constructs are measured by several observable variables, and the relationships between latent variables and indicators are reflective. According to Hair et al. (2012), the reliability and validity must be evaluated for the purpose of interpreting inner model estimates. The standards recommended by Hair et al. (2012) to assess the reflective outer models researches should make fully use of common criteria. In other words, before analyzing the structural model, checking the data reliability and validity is necessary. So a CFA is conducted to establish reflective measures' reliability and validity. The recommended criteria are listed in Table 5.13.

Table 5.13 Recommended Criteria

	Index	Specific Description	Source
Reliability	Internal Consistency Reliability	ronbach's alpha > 0.70; Composite reliability > 0.70	Nunnally and Bernstein, 1994 Fornell and Larcker, 1981
Validity	Convergent Validity	Standardized indicator loadings > 0.70; AVE > 0.50	Fornell and Larcker, 1981
	Discriminant Validity	Fornell-Larcker Criterion, 1981; Cross loadings	

5.3.3 Reliability Analysis

Reliability refers to the extent to which a measure is free of random measurement errors (Smithson, 2005). As an ideal reliable measure, it should not include the random measurement error. On the contrary, most measures employed in the research are not perfect and this situation is not permitted. Therefore, the reliability must be examined to assess the contributions of measure to explain variables (Tharenou et al., 2007).

Although there are many suggestions for evaluating the reliability of each construct (Nunnally, 1978; Hattie, 1985), researchers prefer to focus on Cronbach's alpha coefficient (Cronbach, 1951) which is typically used in social science research to evaluate measurement scale's reliability (Peter, 1979; Cortina, 1993). The alpha coefficient evaluates the closeness of fit for the related items which are included in a group. The sum of the coefficients equals the correlation between scale items. In other words, the average correlation is adopted to calculate alpha coefficient. Therefore, the greater number of items, the larger Cronbach's alpha coefficient will be (Tharenou et al., 2007). Besides, as the average correlation among items increases, the value of Cronbach's alpha also generally increases.

An alpha coefficient ranges from 0 to 1.00. It can be acceptable for exploratory research (Robinson et al., 1991) if the value of alpha is 0.6, even better, if the value larger than 0.70 it can be believed as sufficiently high reliable (Churchill, 1979; Nunnally, 1978; Pallant, 2001).

5.3.4 Validity Analysis

It's argued that if a one-dimensional scale complies with its conceptual definition and satisfies the necessary reliability levels, the instrument's validity should be guaranteed (Hair et al., 2010). Validity is a measure of whether it exactly stands for what it should be. That is, it involves how the concept of interest is defined in the scope of measures. To guarantee the measure of validity, it should begin with an overall consideration of what' s to be measured, and then attempt to make the instrument accurate and precise.

In addition, the validity focuses on the necessity to erase or minimize the impact of the irrelevant factors which can decrease the measure accuracy of the research. The main purpose of instrument validity is to guarantee the accuracy and reality of findings by getting rid of the unrelated variables (Marczyk et al., 2005) .

5.3.5 EFA Results

Exploratory Factor Analysis (EFA) is performed on visual appeal, shopping enjoyment, attitude towards product, and attitude towards store, purchase intention, involvement, privacy concern, issue expertise and information deception detection. Generally speaking, according to Barclay et al. (1995), the measurement models should be confirmed by the following indicators: individual measurement item' s reliability, internal consistency and discriminate validity. The popular method for checking each item' s reliability includes checking whether per item loading is above 0.6 or, perfectly, 0.7 (Chin, 1998; Barclay et al., 1995) . Items' reliability of each construct is measured via the principal components factors analysis method with the varimax rotation for the constructs. Another standard for verifying the instruments adequate discriminant validity should satisfy the following conditions: individual indicator' s loading on their respective latent variables must be higher than the indicator on any other constructs.

Table 5.14 reveals that the measurement items load heavily on their respective constructs and slightly on other factors in our research model. Each item with the loadings is above 0.7, and the value of KMO is 0.872 (the standard level>0.7) . Thus the result confirms adequate convergent and discriminant validity.

Table 5.14 Loadings and Cross-Loadings of Measures

Construct	Item	Component									
		1	2	3	4	5	6	7	8	9	10
PIN	PIN1	.039	.815	.086	.158	.031	.064	.001	.126	−.048	.086
	PIN5	.090	.801	.135	.064	−.024	.105	−.028	.049	.021	.003
	PIN3	.090	.798	.101	.104	.019	.060	−.009	.164	−.086	−.041
	PIN4	.134	.795	.071	.130	.019	.123	−.075	.185	.058	−.073
	PIN2	.105	.775	.163	.061	−.028	.117	−.027	.182	−.022	.030
VA	VA5	.130	.103	.808	.111	−.015	.159	.001	.074	−.044	.036
	VA2	.173	.131	.798	.127	−.023	.033	.006	.160	−.036	−.054
	VA3	.152	.102	.778	.211	.014	.081	−.006	.071	−.065	−.043
	VA4	.134	.129	.775	.176	.041	.100	−.071	.092	.020	.024
	VA1	.100	.123	.773	.157	−.055	.161	−.007	.192	.005	−.062
SE	SE2	.072	.128	.203	.796	−.027	.163	.050	.172	−.049	−.060
	SE4	.170	.114	.217	.770	−.015	.125	−.040	.047	−.054	−.018
	SE3	.085	.112	.161	.748	−.048	.183	−.027	.192	−.033	.023
	SE1	.121	.178	.197	.742	−.009	.092	−.020	.184	.061	−.026
PC	PC2	−.011	−.034	.027	.056	.874	.038	.007	.014	.020	−.065
	PC1	−.062	.011	.008	−.044	.867	−.065	.018	.041	.007	.056
	PC4	−.040	−.002	−.021	−.030	.847	−.044	−.032	−.043	.017	−.029
	PC3	.017	.039	−.042	−.059	.842	.048	.027	.017	−.106	−.026
ATS	ATS1	.328	.130	.160	.169	−.018	.781	.022	.064	−.026	.013
	ATS3	.333	.139	.199	.153	−.026	.767	−.019	.044	.005	−.030
	ATS4	.389	.221	.139	.231	−.010	.760	.040	.048	−.097	.012
	ATS2	.404	.180	.202	.251	.033	.730	.036	.091	−.010	.031
IDD	IDD3	.014	−.065	−.025	−.018	−.063	.020	.867	.073	.047	−.043
	IDD2	−.002	−.045	−.030	.044	.027	.004	.859	−.083	.061	−.006
	IDD1	−.011	.004	−.004	−.054	.053	.012	.853	.017	.003	.020
ATP	ATP1	−.019	.274	.198	.208	.015	.073	.030	.773	.019	.000
	ATP2	.070	.331	.237	.200	.027	.068	−.012	.756	−.119	.001
	ATP3	.055	.290	.236	.283	.002	.050	−.010	.732	.032	.021
IV	IV1	−.044	−.029	−.026	−.046	−.038	−.024	.094	.013	.932	−.006
	IV2	−.077	−.031	−.064	−.010	−.021	−.041	.021	−.052	.930	−.053
EP	EP1	.008	−.030	−.006	−.042	−.043	.011	−.012	.016	−.018	.916
	EP2	−.037	.039	−.064	−.013	−.018	−.002	−.017	−.005	−.038	.910
KMO		0.872									
Bartlett's Test of Sphericity	Approx. Chi-Square	5784.729									
	df	496									
	Sig.	0.000									

Notes: 'PIN' stands for purchase intention. 'VA' stands for visual appeal. 'SE' stands for shopping enjoyment. 'PC' stands for privacy concern. 'ATS' stands for attitude towards store. 'IDD' stands for information deception detection. 'ATP' stands for attitude towards product. 'IV' stands for involvement. 'EP' stands for expertise.

Extraction method: principal component analysis; rotation method: varimax with Kaiser Normalization.

5.3.6 CFA Results

Composite reliability and Cronbach' s alpha values are presented in Table 5.15. As all items' reliability are larger than 0.7, thus the internal consistency standard are satisfied (Nunnally, 1978).

The Cronbach alphas of the measure of ATP, ATS, EP, IDD, IV, PC, PIN, SE, VA are 0.854, 0.866, 0.818, 0.831, 0.872, 0.881, 0.896, 0.871 and 0.893 respectively, indicating adequate reliability of the measurement scales.

Table 5.15 Internal Consistency of Constructs

Construct	Item	Std. path loading	AVE	CR	Cronbach's alpha
ATP	ATP1	0.862	0.774	0.911	0.854
	ATP2	0.896			
	ATP3	0.881			
ATS	ATS1	0.842	0.713	0.909	0.866
	ATS2	0.849			
	ATS3	0.831			
	ATS4	0.854			
EP	EP1	0.882	0.841	0.913	0.818
	EP2	0.950			
IDD	IDD1	0.875	0.744	0.897	0.831
	IDD2	0.878			
	IDD3	0.834			
IV	IV1	0.923	0.884	0.939	0.872
	IV2	0.957			
PC	PC1	0.854	0.500	0.789	0.881
	PC2	0.516			
	PC3	0.486			
	PC4	0.875			
PIN	PIN1	0.855	0.701	0.923	0.896
	PIN2	0.813			
	PIN3	0.850			
	PIN4	0.844			
	PIN5	0.836			
SE	SE1	0.8330	0.722	0.912	0.871
	SE2	0.884			
	SE3	0.835			
	SE4	0.847			
VA	VA1	0.825	0.700	0.921	0.893
	VA2	0.837			
	VA3	0.838			
	VA4	0.842			
	VA5	0.841			

Note: CR represents composite reliability; AVE represents Average Variance Extracted.

Usually, to evaluate the measurement model requires checking its discriminant validity. As listed in Table 5.16, off-diagonal elements reveal all the latent variables' correlations, while the diagonal elements are the square roots of the average variances extracted (AVE) of the latent variables. In line with the study of Barclay et al. (1995), for the adequate discriminant validity, any latent variable's average variances extracted (AVE) should be larger than the variance which is shared between the latent variable and other latent variables. In another word, the diagonal elements should be larger than corresponding off-diagonal elements. Therefore, the data presented in Table 5.16 satisfy this requirement.

To further eliminate the negative effects from common method variance (CMV), following the previous studies (e.g., Malhotra et al., 2006), the study creates a marker variable, privacy concern, which is taken as the indicator of all endogenous variables: visual appeal, shopping enjoyment, attitude towards product, attitude towards store and purchase intention. For each endogenous variable, pseudo-test is conducted to check whether the improved R^2 is significant or not (Subramani, 2004). The previous study argues that if the coefficient is under 0.10 at $\alpha=0.05$ levels, then the common method variance (CMV) is not significant (Malhotra et al., 2006). In Table 5.16, the results suggest that the original correlations among attitude towards product, attitude towards store and other variables do not differ significantly ($\Delta r=-0.018$). Hence, it is concluded that the path coefficients in this research model are not influenced by common method variance, so CMV is not a potential threat.

Table 5.16 Correlation Matrix

	ATP	ATS	EP	IDD	IV	LY	PC	PIN	SE	VA
ATP	0.879									
ATS	0.325	0.844								
EP	-0.004	-0.001	0.917							
IDD	-0.021	-0.017	-0.024	0.863						

续表

	ATP	ATS	EP	IDD	IV	LY	PC	PIN	SE	VA
IV	-0.085	-0.019	-0.063	0.097	0.940					
LY	0.156	0.627	-0.034	0.009	-0.012	0.861				
PC	-0.013	-0.125	0.011	-0.005	0.014	-0.112	0.707			
PIN	0.578	0.397	0.004	-0.074	-0.077	0.241	-0.020	0.840		
SE	0.517	0.499	-0.057	-0.030	-0.081	0.283	-0.107	0.357	0.849	
VA	0.472	0.452	-0.069	-0.046	-0.103	0.336	-0.058	0.339	0.477	0.837

Note: Diagonal elements are the square root of AVE. The value should exceed the inter-construct correlations for adequate discriminant validity.

5.4 Structural Model

5.4.1 Structural Equation Modeling

Structural equation modeling involves a diverse analysis method which includes path analysis, confirmatory factor analysis, multiple regression analysis and partial least squares path analysis to evaluate simultaneously a battery of interrelated dependency (Hair et al., 1998). It is developed from the models in econometrics and its main application has been increasing in psychology and social sciences. As an example, SEM is commonly used in the field of attitude theory (Perugini and Conner, 2000; Shim and Eastlick, 1998; Hansen et al., 2004). Its popularity is adopted in the usage of confirmatory methods which afford a systemic approach for evaluating and improving theoretical models for researchers.

The significant advantage of SEM is to model the relationships between latent variables and observed variables. Latent variables refer to those that cannot be observed but can be administered via the process of scale development. Put another way, there is a bridge between latent variables and factors in doing factor analysis; SEM can be taken as a compound of factor analysis and regression or path analysis. A SEM model usually includes two parts: measurement model and structural model. The measurement model focuses on the relationship between latent variable, its (observed) measures and their measurement errors.

The measurement model is formed by performing confirmatory factor analysis (CFA) which is adopted to acquire a robust construct. According to Anderson and Gerbing (1988), all the scales need to be tested through the estimation of the measurement model. The factor structure (the observed variables mapping on to the latent factors) is then used to estimate the full latent variable structural path model (Kelloway, 1996). With regard to the structural model, both the correlation relationships and dependence relationships between variables should include the core factors of the model (Hair et al., 2010).

5.4.2 Partial Least Squares (PLS)

Partial Least Squares (PLS) is a soft modeling approach to SEM with no assumptions about data distribution (Vinzi et al., 2010). Partial least squares permit the multivariate analysis technique using the simultaneous analysis strategy for predictor variables and various criteria (Barclay et al., 1995). PLS approach overcomes some problems associated with correlation analyses and appropriates for explorative analysis towards structural equation models. According to Wold (1985), PLS is a strong analysis method because of its minimal demands on measurement scales, residual distributions and sample size. PLS-SEM has been deployed in many fields, such as behavioral sciences (e.g., Bass et al., 2003), marketing (e.g., Henseler et al., 2009), organization (e. g., Sosik et al., 2009), business strategy (e. g., Hulland, 1999) and management information system (e.g., Chin et al., 2003).

PLS path modeling focuses on the prediction of the dependent variables (both latent and manifest). This objective is achieved by maximizing the explained variance (R^2) of the dependent variables. Thus, parameter estimates are obtained based on the ability to minimize the residual variances of dependent variables. Compared to covariance-based SEM, PLS is more suited for predictive applications and theory building (exploratory analysis), although PLS can be also used for theory confirmation (confirmatory analysis). Chin (1998b, p. 299) makes three basic distinctions for choosing between PLS and CBSEM: (1) the constructs are modeled as indeterminate or determinate (defined); (2) the extent to which, in the theoretical model and auxiliary theory, the researcher links measures to constructs, and (3) the researcher is parameter-oriented or prediction-oriented. If the second option applies best to each question, the PLS approach is likely to be more suitable. Certainly, PLS can be a powerful method of analysis because of the minimal demands on measurement scales, sample size, and residual distributions (Fornell and Bookstein, 1982). With reference to CBSEM, PLS avoids two serious problems which often interfere with meaningful modeling: improper, i. e. inadmissible solutions (Fornell and Bookstein, 1982), and factor indeterminacy.

PLS is a technique designed to reflect the theoretical and empirical conditions present in the behavior of a land social science, where these are habitual situations with no solid theories and scarce knowledge, and this kind of modeling is called soft modeling (Wold, 1980). A PLS model includes two components: the first part is structure equation model, which represents the relationship among constructs, and the second part is measurement model, which represents the relationship between constructs and their respective indicators (Fornell and Bookstein, 1982). According to Wold (1985), in partial least squares, the constructs' indicators could be identified into two types: reflective indicator or formative indicator. Reflective indicators are estimated via the constructs they represent, hence, reflective indicators covary with its own construct levels (Chin and Gopal, 1995). In contrast, the formative indicators state the construct they represent, and it is possible for formative indicators not to vary with the construct they determined to some extent (Chin and Gopal, 1995). As all constructs' indicators are expected to covary with the levels of their respective construct, therefore, all those constructs' indicators are modeled reflectively.

In PLS analysis, Gefen et al. (2000) and Chin (1998) suggest there are two evaluation criteria: firstly, the size of sample should be at least 10 times the number of items attached to the most complex construct. The second criterion is that at least 10 times the largest number of paths is directed from independent variable to a dependent variable. In this study, the total sample size of 331 is more than sufficient to run PLS estimation.

After the measure model is verified followed by evaluating structural model. Hair et al. (2013) propose the following standards for evaluating the structural model: path coefficients, R^2 and t-value. R^2 represents a construct's total explanation variance by its antecedent variables in a specific research model (Lee et al., 2011). Path coefficients and t-value are used to reveal the relationships between two variables. According to Thompson et al. (1995), the higher R^2 and larger t-values can indicate the significant effect between variables and thus can accept the hypothesis.

Based on discussion above, this study adopts the partial least squares to test consumers' affective perception towards attitudes and purchase intention.

5.4.3 Step Two: PLS Test for Effective Responses towards Attitudes and Behavior

Based on the discussion above, in the second step, this study attempts to perform PLS test to validate the consumers' effective perception towards attitudes and purchase intention. The second step of validating research model is shown in Figure 5.4.

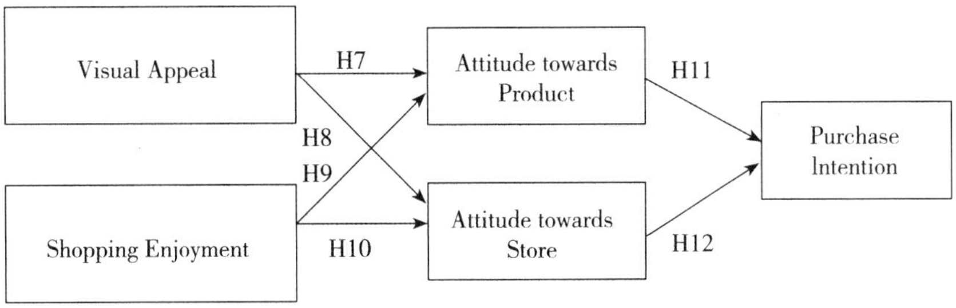

Figure 5.4 The Second Step of Validating Research Model

The PLS test outcomes display that all causal paths are statistically significant ($p<0.001$) for visual appeal to attitude towards product, visual appeal to attitude towards store, shopping enjoyment to attitude towards product, shopping enjoyment to attitude towards store, attitude towards product to purchase intention. The structural model's explanatory power is explained by the value of R^2 of the dependent variables. To verify the hypotheses, the t statistics is calculated to gather the standardized path coefficient, and then the p-values based on the two-tailed test and significant level at 0.05 are examined respectively. The outcomes of the PLS analysis are presented in Figure 5.5.

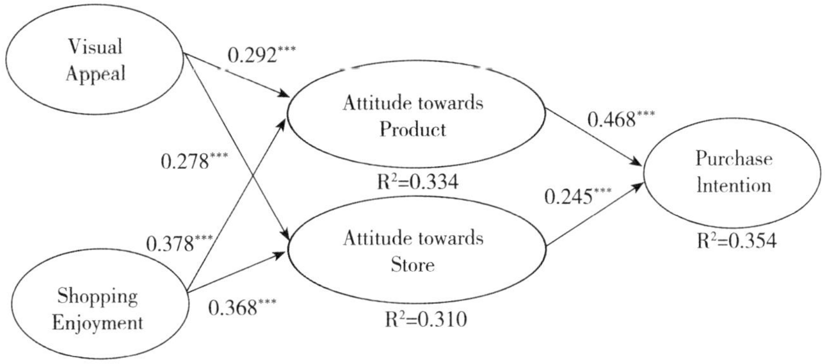

Figure 5.5 PLS Analysis Results

First of all, the researcher assesses the predictive power of the model by examining the R^2 values on the endogenous variables. As can be seen in Figure 5.5, the R^2 of purchase intention equals 0.354, which shows that

35.4% of purchase intention may be explained by attitude towards product and attitude towards store. Both attitude towards product （$p<0.001$） and attitude towards store （$p<0.001$） have significant impact on purchase intention. The R^2 of attitude towards product is 0.334 which means that 33.4% attitude towards product can be explained by visual appeal and shopping enjoyment. The $R^2$of attitude towards store is 0.310 which means that 31.0% attitude towards store can be explained by visual appeal and shopping enjoyment. Path coefficients β， t-value and standard error could be seen in Table 5.17.

Table 5.17 Path Coefficients

Hypothesis		Path Coefficients β	T-value	Standard Error	Support
H7	VA -> ATP	0.292^{***}	6.108	0.0485	S
H8	VA -> ATS	0.278^{***}	5.155	0.0539	S
H9	SE->ATP	0.378^{***}	8.027	0.0472	S
H10	SE -> ATS	0.368^{***}	6.637	0.0554	S
H11	ATP -> PI	0.468^{***}	11.429	0.0409	S
H12	ATS -> PI	0.245^{***}	5.273	0.0466	S

Note: T-values greater than 1.96 are significant at $*p<0.05$， those greater than 2.576 are significant at $**p<0.01$， and those greater than 3.291 are significant at $***p<0.001$. ‘S’ stands for supported.

Visual appeal （β=0.292， t-value=6.108， $p<0.001$） has significant effect on attitude towards product， so hypothesis H7 is supported. Visual appeal （β =0.278， t-value=5.155， $p<0.001$） has significant effect on attitude towards store， so hypothesis H8 is supported. Shopping enjoyment （β =0.378， t-value=8.027， $p<0.001$） has significant effect on attitude towards product， so hypothesis H9 is supported. Shopping enjoyment （β= 0.368， t-value=6.637， $p<0.001$） has significant effect on attitude towards store， so hypothesis H10 is supported. Attitude towards product （β=0.468， t-value=11.429， $p<0.001$） has significant effect on purchase intention， so hypothesis H11 is supported. Attitude towards store （β =0.245， t-value=

5.273, $p<0.001$) has significant effect on purchase intention, so hypothesis H12 is supported.

All hypotheses (H7~H12) are approved statistically significant. Such R^2 values of the dependent variables are calculated as attitude towards product (0.334), attitude towards store (0.310) and purchase intention (0.354). Based on the bootstrap resampling analysis, the path significant levels are performed on the structural model.

Overall, based on the discussion above, in this study, ANOVA is used to empirically test the left-hand of research model and the PLS algorism analyses test the right-hand of research model and related hypotheses. The analysis results suggest that photo decoration cues can significantly influence consumers' affective reactions, and the affective reactions improve consumers' attitude towards grocery product and attitude towards store, which jointly form the decisive factors of purchase intention. The results of hypotheses testing are summarized in Table 5.18.

Table 5.18 Summary of Hypotheses Test

No.	Hypotheses	Result
H1	Reflective surface positively affects online consumers' visual appeal.	Supported
H2	Contrast color usage positively affects online consumers' visual appeal.	Supported
H3	Complementary goods layout positively affect online consumers' visual appeal.	Supported
H4	Reflective surface positively affects online consumers' shopping enjoyment.	Supported
H5	Contrast color usage positively affects online consumers' shopping enjoyment.	Not Supported
H6	Complementary goods layout positively affect online consumers' shopping enjoyment.	Supported
H7	Visual appeal positively affects consumers' attitude towards products.	Supported

续表

No.	Hypotheses	Result
H8	Visual appeal positively affects consumers' attitude towards online grocery store.	Supported
H9	Shopping enjoyment positively affects consumers' attitude towards product.	Supported
H10	Shopping enjoyment positively affects consumers' attitude toward online grocery store.	Supported
H11	Consumers' attitude towards product positively affects their purchase intention.	Supported
H12	Consumers' attitudes towards grocery store positively influence their purchase intentions.	Supported

Chapter 6 Conclusion

6.1 Overall Conclusion

This study seeks to answer two research questions: what photo decoration cues in online grocery could be identified? Could photo decoration cues significantly affect the affective perception and behavioral intention of consumers? Using the media richness theory, visual rhetoric theory and visual design literature as the theoretical foundation, the study provides a solid foundation to comprehend the impacts of photo visual design artifacts on consumers' affective responses when online consumers shop online for grocery, in other words, what is consumers' focus of attention towards a grocery product's photo. In this study, the reflective surface, contrast color usage and complementary goods layout as stimuli of consumers' affective reactions are conceptualized and empirically tested.

For the purpose of research, this study includes six chapters. Chapter 1 provides a research context, purpose of research and research questions. Chapter 2 describes theoretical background about media richness theory, visual rhetoric theory, and visual design literature, attitude towards product and attitude towards store which backup this study. In Chapter 3, hypotheses are developed followed by a research model which is adopted to validate how photo decoration cues affect consumers' behavior through the effective responses, attitude towards grocery store and attitude towards products. Chapter 4 presents the research methodology and laboratory experiment design procedure. Chapter 5 analyzes the data to test the research model and hypotheses are also developed in this part. Chapter 6 addresses the conclusion, implications, contributions, research limitations and future research direction.

Using the data based on questionnaire survey, the suggested research model examines the relationships among the different states of photo decoration cues and consumers' effective responses (visual appeal and shopping enjoyment). It is predicted that photo decoration cues will induce consumers' affective responses in which the grocery product photo will appear more appealing. In turn, affective responses are predicted to result in purchase intention. The findings provide strong support for the proposed model, as expected, photo decoration cues are associated with consumers' affective actions, attitudes towards product and store which in turn positively affect consumes' purchase intention.

All expected relationships are supported in the research model, except Hypothesis 5. The findings show that contrast color usage does not significantly affect consumer shopping enjoyment. This is different from the initial expectations based on the literature. There are three plausible explanations for this finding. First, the participants may have gathered the sufficient and critical information they need which elicits shopping enjoyment from other decorations cues of the photo, such as reflective surface and complementary goods layout, leading them to perceive few benefits and little relevance towards contrast color usage. The second reason comes from the experiment stimulus, where the white or red plate chosen in the experiment is relatively simple, which means that the plate's color has fewer characteristics, as a result, participants cannot be stimulated enough to make their shopping enjoyment. The last reason contributes to the culture differences among countries. This is because that the culture difference also plays a role to influence customer's judgment, reaction and evaluation results. For example, different countries have different meanings for the same color.

The findings indicate that the photo decoration cues are the important constructs and serve as the antecedents of consumers' affective responses. Therefore, an understanding of how to design photo artifacts which can influence consumers' affective responses will be very important. In addition, this study has demonstrated that the visual appeal and shopping enjoyment evoked by affective responses have indeed a positive impact on the attitudes towards product and store. This implies that the more decoration cues are used in designing online grocery photo, the more likely they will foster a positive attitude towards using them.

The findings of this study will increase the ever-increasing body of literature about online product presentation and will also serve as a reference for vendors to improve the design of their grocery website. Thus, by providing photo visual decoration cues, website vendors can effectively stimulate consumers' understanding about products.

6.2 Contribution

This study provides an innovative perspective on designing grocery presentations which focuses on the visual decoration cues in grocery photo presentation and reveals the stimulus of three different cues on consumers' affective reactions. Specifically, this research highlights the photo decoration cues as the affective stimulus to provoke consumers' affective perception. Thus, it is necessary for online vendors to adopt appropriate product presentation technology to draw customers' attention and evoke their affective reaction to visit website.

The findings insist that the innovative product presentation that creates visual appeal and shopping enjoyment will lead to increased purchase intentions. Thus, with the unique focus on grocery decoration cues and its impact on consumers' affective reactions, this study presents a new theoretical standpoint for information systems and marketing research fields.

This research develops a theoretical model with the core variables influencing consumers' affective response which is the first attempt to extend media richness theory and visual rhetoric theory in consumers' affective perception and apply the findings to the background of online grocery presentation.

6.2.1 Contribution to Academics

As there are fewer empirical studies to investigate whether and how photo decoration cues in online grocery are associated with customers' affective responses, perceptions and attitudes, the findings of this study have important contributions from a theoretical perspective.

First, by controlling possible extraneous variables in a laboratory experiment, the findings confirm the existence of the photo decoration cues in the grocery photo design and empirically test the impact of photo decoration cues on consumers' affective responses.

Second, in the literature on information system and marketing, the dominant theoretical framework to study product presentation are virtual reality, media choice and interactive function paradigm, whereas in this study, it is argued that photo design artifacts influence the level of customers' affective reaction, which in turn determines customers' purchase decisions. This might be the first study that empirically explains photo decoration cues on customers' affective responses in e-commerce context. Consequently, this study contributes to the literature of e-commerce on product presentation.

As online grocery market has been growing quickly over the past few years, however, a limited number of researches focused on the online grocery market, especially very limited studies have addressed decoration cues in online grocery presentation. Hence, this dissertation can contribute to the existing literatures on the product presentation in the grocery market.

Third, the current study combines the Media Richness Theory and Visual Rhetoric Theory to reach the aim of persuasion to stimulate customers' affective responses. The results indicate the sufficient feasibilities of adopting the literature of visual design and applying the theories of media richness and visual rhetoric to conduct the research in the e-commerce domain.

This research has also contributed to the extendedness of visual design literature. Because visual design has the potential to arouse consumers' emotional appeal (Garrett, 2003), and could be expressed via photos or colors (Rosen and Purinton, 2004). This study has explored the impact of photo design artifacts that impart the consumers' affective reactions, which creates a new theoretical perspective that moves beyond the TAM (Technology Acceptance Model) into the range of affective reactions.

In addition, from the human-computer interaction (HCI) perspective, this study merges the research fields of human-computer interaction and e-commerce to understand the impact of photo design artifacts on consumer affective responses when consumers shop online for grocery. In answer to the call for human-computer interaction research (Zhang et al., 2002; Carey et al., 2004), this study sets up a bridge between human-computer interaction and information systems (IS) by examining the details of photo design artifacts of online grocery products' presentation. Most previous researches have discussed the product presentation methods just by focusing on comparing them with other optional interface conditions (an experimental research method that is valid for assessing a technology), but rarely sheds light on the photo design aspects especially for the photo decoration cues. For example, even though many previous researches have discovered that VPE (virtual product experience) technology greatly benefits consumers and users, they do not distinguish the core elements of design aspects towards online product presentations (Suh and Lee, 2005; Li et al., 2002, 2003) or do not elaborate how each design aspect works and what the relative contributions of different design are (Jiang and Benbasat, 2005), although designers and researchers normally intend to clearly know about how a technology works and its direct impacts (Hevner et al., 2004).

Furthermore, this research makes a significant contribution to the literature by extending the knowledge of online purchase intention beyond the much studied influence of product presentation, assessing the influence of photo decoration cues and their role as triggers for the adoption of online grocery shopping. This study also contributes to how the visual rhetoric theory is applied to users' perceptions of photo decoration cues in grocery website design. According to Salinas (2002, p.168), photos are the "strategic components of visual representations" which afford meanings to the observers. Our findings suggest that the photo decoration cues' conveyance ability to present product and deliver information has potential to attract consumers' attention to the purchase task. Therefore, comparing with the media selection, the cues' conveyance ability is more likely to catch consumers' attention.

Finally, our study fills the gaps between the absence of rigorous empirical research on the photo design artifacts in the IS and marketing fields and the wide usage of photo on the website.

6.2.2 Contribution to Practices

E-commerce designers have long been paid a close attention to the design issues of website. Despite the suggestion provided by various studies, problems in website design continue to a low progress in practice because of the fact that few existing researches have tried to put forward how specific aspects of photo decoration artifacts may affect the success of e-commerce website. It is imperative for managers and designers to understand that photo decoration cues should be carried out the direct aim of increasing consumers' affective responses, which will allow online retailers to be more successful to provoke positive outcomes in the design of e-commerce website.

Specifically, it is very interesting for advertisers and online vendors when they expect their product presentations can effectively deliver product information to users. "One of the most critical attention issues in electronic commerce is simply attracting it" (Davenport and Beck, 2001). It's reported that more and more online vendors are providing the new job positions, such as cognitive designers and affective designers, to validate the relationship among consumers' behavior, attention and interface design. The findings of this study can benefit website designers by handling the general beliefs of employing photo decoration cues, and putting forward a more comprehensive and precise understanding of its effects on consumers' affective reactions and perceptions. At the same time, retailers should consider and apply the optimal level of decoration cues in their products' photo to facilitate consumers' understanding during their online shopping activities.

Furthermore, one of the main research trends of human-computer interface research is developing the user-centered systems, especially for the topic of interface design. An interface design with the photo decoration cues technologies which attract consumers' attention can act as technology leverage point to improve product performance. In this study, the effects of photo design artifacts on consumers' reaction and perceptions are verified via Media Richness Theory and Visual Rhetoric Theory. Therefore, there is a great potential to apply theories and research findings from human-computer interface to IS and marketing research areas.

This study provides useful guidelines for online vendors to implement photo decoration cues to advertise their product. If web-based grocery stores want to enhance their product presentation with more convenient concerns for consumer, they can achieve it by adopting photo decoration cues for grocery, because implementing photo decoration cues is relatively cheaper in cost than other technologies.

Exploring the design artifacts of photo for the commercial rhetoric, this study finds a rich path for the research of consumers' behaviors science and may dedicate to the literature of design and persuasion in other research fields. Thus, the beneficiaries of this research work may be not only consumers, but also photographers and designers. In addition to calling for the new research paths, a rhetorical method can help in formulating answers to advertisement maker. Visual rhetoric research should investigate how visual signs are handled as well as how photos are manufactured, adopted and interpreted in advertisement area.

The findings also indicate that the strong attention attracted by interface features, such as photo decoration artifacts, can help online users identify the characteristics of grocery quickly in product presentations.

6.3 Limitations and Future Research

Although this study provides some meaningful results for research and practice, there are some limitations.

First, this study utilizes student participants partly, which is usually blamed for lacking generalizability. Although this kind of student consumer group may be taken for a suitable participant group because of their convenience with technology and online shopping context, it confines the external validity of the study. Replication of this study with a larger and more representative sample will increase the generalizability of the results. For the phenomenon examined in this study, this limitation is not a major issue. Future research should investigate different category of participants with more various demographic characteristics to validate the findings.

Second, the study employs a relative simple and easy experimental website as well as the experimental task requirement. Although the experimental task has been designed to induce a real shopping scenario on the grocery store, it may still appear not natural. As mentioned above, to guarantee higher internal validity, it's necessary for some control methods to be adopted in the experiment. To achieve the accurate laboratory outcomes, in fact, to control manipulation and eliminate possible "contaminating" situations are possible (Kerlinger et al., 1999). If higher external validity is required, it's necessary to implement a field study at the cost of decreased internal validity and lower accuracy of the results.

Third, the photo design is not perfect although some special photographic technologies are adopted. The adoption of contrast color between the target product and background is unsatisfactory, in other words, maybe the plates' color is not ideal. In addition, the whole effects are not so perfect.

Fourth, the research model is tested in China grocery shopping background, while various studies have insisted that the culture differences should play a role to influence customer's judgment, reaction and evaluation results. The fitness to other kinds of product presentation is not research yet. Further research should be generalized to other culture background.

Fifth, most hypotheses are validated via a laboratory experiment which benefits a lot to interpret the relationship between various variables more stringently. However, the real word is not changeless; a laboratory experiment may not response to the real world dynamically. To increase external validity, future works should employ different methods such as survey and interview which have advantages to reflect the dynamical world.

This study opens various channels for the future research. First, while photo decoration cues are the focus in this study, it would be useful to examine the effects of other forms of photo design artifacts, such as coordinate effects between main product and complementary goods. Future research should conceptualize other important photo decoration cues and develop new scales for them. Second, the effects of photo decoration cues are examined in a grocery shopping environment, online browsing and other major online tasks that users participate in. It would be interesting to duplicate these research findings to the browsing and searching context for positioning the targets by the decoration cues rapidly. Third, other interface effects, such as the background color, can be examined together with product' s color to study the coordinate effect and contrast effect. In addition, there may be some relationships that exist between visual appeal and shopping enjoyment, attitude towards product and attitude towards store. All those are needed to conduct in the future research.

Future research could be conducted to explore the differences between consumers both with and without experience of online grocery shopping, to get more thorough understanding about their perceptions. Also researchers should further investigate the effects of reflective surface and complementary goods layout with regard to visual aspects, because contrast color usage reflects visual appeal more sufficiently than the focus on shopping enjoyment in this study.

This study explores a new linkage between the theory of media richness and visual rhetoric, thus, future research could be based on their integrated theoretical underpinning to dig some new implications in a deeper level.

REFERENCES

[1] Abernethy, A. M. and Franke, G.R. (1996), "The information content of advertising: A meta-analysis," Journal of Advertising, Vol.25, No.2, pp.1-17.

[2] Agarwal, R. and Karahanna, E. (2000), "Time flies when you're having fun: Cognitive absorption and beliefs about information technology usage," MIS Quarterly, pp.665-694.

[3] Ajzen, I. (1985), "From intentions to actions: A theory of planned behavior," Action Control from Cognition to Behavior. Springer, pp.11-39.

[4] Ajzen, I. (2001), "Nature and operation of attitudes," Annual Review of Psychology, Vol.52, No.1, pp.27-58.

[5] Anderson, R. L. and Ortinau, D. J. (1988), "Exploring consumers' post adoption attitudes and use behaviors in monitoring the diffusion of a technology-based discontinuous innovation," Journal of Business Research, Vol.17, No.3, pp.283-298.

[6] Andrews, J. C. and Shimp, T. A. (1990), "Effects of involvement, argument strength, and source characteristics on central and peripheral processing of advertising," Psychology & Marketing, Vol.7, No.3, pp.195-214.

[7] Areni, C. S. and Kim, D. (1994), "The influence of in-store lighting on consumers' examination of merchandise in a wine store," International Jour-

nal of Research in Marketing,Vol.11,No.2,pp.117-125.

[8] Argyle, M., Alkema, F. and Gilmour, R.(1971), "The communication of friendly and hostile attitudes by verbal and non-verbal signals," European Journal of Social Psychology ,Vol.1,No.3,pp.385-402.

[9] Avery,J.,Steenburgh,T. J.,Deighton,J.(2012),"Adding bricks to clicks: Predicting the patterns of cross-channel elasticities over time," Journal of Marketing ,Vol.76,No.3,pp.96-111.

[10] Aylott, R. and Mitchell, V. W.(1998), "An exploratory study of grocery shopping stressors," International Journal of Retail & Distribution Management,Vol.26,No.9,pp.362-373.

[11] Babakus, E. and Boller, G. W.(1992), "An empirical assessment of the SERVQUAL scale," Journal of Business research,Vol.24,No.3,pp.253-268.

[12] Babin,B. J.,Darden,W. R. and Babin,L. A.(1998a),"Negative emotions in marketing research: affect or artifact? " Journal of Business Research Vol.42,No.3,pp.271-285.

[13] Babin,B. J.,Darden,W. R. and Babin,L. A.(1998b),"Negative emotions in marketing research:affect or artifact? ",Journal of Business Research, Vol.42,No.3,pp.271-285.

[14] Babin,B. J.,Darden,W. R. and Griffin,M.(1994),"Work and/or fun: measuring hedonic and utilitarian shopping value," Journal of Consumer Research,pp.644-656.

[15] Baggett,P.(1989),"Understanding Visual and Verbal Messages," Advances in Psychology,Vol.58,pp.101-124.

[16] Banerjee,S.,Bhattacharjee,B. and Kommareddy,C.(2002),Scalable Application Layer Multicast.

[17] Barclay, D., Higgins, C. and Thompson, R.(1995). "The partial least squares(PLS) approach to causal modeling: Personal computer adoption and use as an illustration," Technology Studies,Vol.2,No.2,pp.85-309.

[18] Bass,B. M.,Avolio,B. J.,Jung,D. I. and Berson,Y.(2003), "Predicting unit performance by assessing transformational and transactional leadership," Journal of Applied Psychology,Vol.88,No.2,pp.207-218.

[19] Benn,Y.,Webb,T. L.Chang,B. P.(2015),"What information do consumers consider,and how do they look for it,when shopping for groceries online? " Appetite,Vol.89,Jun,pp.265-273.

[20] Besuijen, K. and Spenkelink, G. P.(1998), "Standardizing visual display

quality," Displays, Vol.19, No.2, pp.67-76.

[21] Bodur, H. O., Brinberg, D. and Coupey, E. (2000), "Belief, affect, and attitude: Alternative models of the determinants of attitude," Journal of Consumer Psychology, Vol.9, No.1, pp.17-28.

[22] Brady, L. and Phillips, C. (2003), "Aesthetics and usability: A look at color and balance," Usability News, Vol.5, No.1, pp.1-4.

[23] Brown, S. P., Homer, P. M. and Inman, J. J. (1998a), "A meta-analysis of relationships between ad-evoked feelings and advertising responses," Journal of Marketing Research, pp.114-126.

[24] Brown, S. P., Homer, P. M. and Inman, J. J. (1998b), "A meta-analysis of relationships between ad-evoked feelings and advertising responses", Journal of Marketing Research, pp.114-126.

[25] Budimir, S. and Palmović, M. (2011), "Gaze differences in processing pictures with emotional content," Collegium Antropologicum, Vol.35, No.1, pp.17-23.

[26] Burke, K. (1969), A Rhetoric of Motives, University of California Press.

[27] Burke, R. (2002), "Technology and the customer interface: what consumers want in the physical and virtual store," Journal of the Academy of Marketing Science, Vol.30, No.4, pp.411-432.

[28] Cai, S. and Xu, Y. (2007), "Effects of outcome, process and shopping enjoyment on online consumer behavior," Electronic Commerce Research and Applications, Vol.5, No.4, pp.272-281.

[29] Campbell, D. T. and Stanley, J. C. (1963), "Experimental and quasi-experimental designs for research on teaching," American Educational Research Association.

[30] Carey, J., Galletta, D. F.Kim, Y. (2004), "The role of human computer interaction in management information systems curricula: A call to action," The Communications of the Association for Information Systems, Vol.13, No.1, p. 62.

[31] Chau. PYK, A. G., Tam. KY (2000), "Impact of information presentation modes on online shopping an empirical evaluation of a broadband interactive shopping service", Journal of Organizational Computing And Electronic Commerce, Vol.10, No.1, pp.1-20.

[32] Chien, A. W., S. and von Westarp, F. (2003), "The acceptance of online grocery shopping," BLED 2003 Proceedings, p.52.

[33] Childers, T. L., Carr, C. L. Peck, J. (2002), "Hedonic and utilitarian motivations for online retail shopping behavior," Journal of Retailing, Vol. 77, No.4, pp.511-535.

[34] Chin, W. (1998), "The partial least squares approach to structural equation modeling," Modern Methods for Business Research, Vol. 295, No. 2, pp.295-336.

[35] Chin, W. and Gopal, A. (1995), "Adoption intention in GSS: relative importance of beliefs," ACM SigMIS Database, Vol.26, No.2-3, pp.42-64.

[36] Chin, W. W. (1998), "the partial least squares approach to structural equation modeling," Modern Methods for Business Research, Vol.295, No.2, pp.295-336.

[37] Chin, W. W., Marcolin, B. L. and Newsted, P. R (2003), "A partial least squares latent variable modeling approach for measuring interaction effects: Results from a monte carlo simulation study and an electronic-mail emotion/adoption study," Information Systems Research, Vol. 14, No. 2, pp.189-217.

[38] Chou, K. C. (2011), "Some remarks on protein attribute prediction and pseudo amino acid composition," Journal of Theoretical Biology, Vol.273, No.1, pp.236-247.

[39] Christensen, C. M. and Tedlow, R. S. (2000), "Patterns of disruption in retailing," Harvard Business Review, Vol.78, No.1, pp: 42 45.

[40] Churchill Jr, G. A. (1979), "A paradigm for developing better measures of marketing constructs," Journal of Marketing Research, pp.64-73.

[41] Corbett, E. (1965), Classical Rhetoric for the Modern Student, London: Oxford.

[42] Corral, C. B. (1999), "On-line grocery shopping heats up despite concern over big issues," Discount Store News, Vol.8, No.13, pp.18-20.

[43] Coyle, J. R. and Thorson, E. (2001), "The effects of progressive levels of interactivity and vividness in web marketing sites," Journal of Advertising, Vol.30, No.3, pp.65-77.

[44] Cry, D. Head, M., Larios, H. (2009), "Exploring human images in website design: A multi-method approach", MIS Quarterly, Vol. 33, No. 3, pp.539-566.

[45] Csikszentmihalyi, M. (1975), Beyond Boredom and Anxiety, San Fransisco: Jossy-Bass.

[46] Csikszentmihalyi, M. (1990), Flow: The Psychology of Optimal Experience, New York: Harper Perennial.

[47] Csikszentmihalyi, M., Larson, R. and Prescott, S. (1977), "The ecology of adolescent activity and experience," Journal of Youth and Adolescence, Vol.6, No.3, pp.281–294.

[48] Cudeck, R. and O´Dell, L. L. (1994), "Applications of standard error estimates in unrestricted factor analysis: significance tests for factor loadings and correlations," Psychological Bulletin, Vol.115, No.3, p.475.

[49] Cyr, D. (2008), "Modeling website design across cultures: Relationships to trust, satisfaction, and e–loyalty," Journal of Management Information Systems, Vol.24, No.4, pp.47–72.

[50] Cyr, D., Bonanni, C.Bowes, J. (2005), "Beyond trust: Website design preferences across cultures," Journal of Global Information Management (JGIM) Vol.13, No.4, pp.25–54.

[51] Cyr, D., M. and Ivanov, A. (2006), "Design aesthetics leading to m–loyalty in mobile commerce," Information & Management, Vol.43, No.8, pp.950–963.

[52] Dabholkar, P. A. and Bagozzi, R. P. (2002), "An attitudinal model of technology–based self–service: moderating effects of consumer traits and situational factors," Journal of the Academy of Marketing Science, Vol.30, No.3, pp.184–201.

[53] Daft, R. L. and Lengel, R. H. (1986), "Organizational information requiremonts, media richness and structural design," Management Science, Vol.32, No.5, pp.554–571.

[54] Daft, R. L., Lengel, R. H. and Trevino, L. K. (1987), "Message equivocality, media selection, and manager performance: Implications for Information Systems," MIS Quarterly, pp.355–366.

[55] Davenport, T. and Beck, J. (2001), "The attention economy: Understanding the new economy of business," Cambridge, MA: Harvard Business School Press.

[56] Davis, F. D. (1989), "Perceived usefulness, perceived ease of use, and user acceptance of information technology," MIS Quarterly, pp.319–340.

[57] Davis, F. D. (1993), "User acceptance of information technology: system characteristics, user perceptions and behavioral impacts," International journal of Man–machine Studies, Vol.38, No.3, pp.475–487.

[58] De Vaus, D. (2002), Analyzing Social Science Data: 50 Key Problems in

Data Analysis. Sage Publications.

[59] Dekimpe, M. G., Steenkamp, J. B. E. Mellens, M. (1997), "Decline and variability in brand loyalty," International Journal of Research in Marketing, Vol.14, No.5, pp.405-420.

[60] Demangeot, C. and Broderick, A. J. (2010), "Consumer perceptions of online shopping environments: A gestalt approach," Psychology & Marketing, Vol.27, No.2, pp.117-140.

[61] Dennis, A. R. and Kinney, S. T. (1998), "Testing media richness theory in the new media: the effects of Cues, Feedback, and Equivocality," Information Systems Research, Vol.9, No.3.

[62] DeSanctis, G. (1984), "Computer graphics as decision aids: directions for research", Decision Sciences, Vol.15, No.4, pp.463-487.

[63] DeVellis, R. (1991), Scale Development: Theory and Applications: Sage, Newbury Park, CA.

[64] Dick, A. S. and Basu, K. (1994), "Customer loyalty: towards an integrated conceptual framework," Journal of the Academy of Marketing Science, Vol.22, No.2, pp.99-113.

[65] Duncan, J. and Humphreys, G. W. (1989), "Visual search and stimulus similarity," Psychological Review, Vol.96, No.3, pp.433.

[66] Dwyer, F. R., Schurr, P. H. and Oh, S. 1987. "Developing buyer-seller relationships," The Journal of Marketing, pp.11-27.

[67] Eagly, A. H., Mladinic, A. and Otto, S. (1994), "Cognitive and affective bases of attitudes towards social groups and social policies," Journal of Experimental Social Psychology, Vol.30, No.2, pp.113-137.

[68] Eastlick, M. A. and Lotz, S. (1999), "Profiling potential adopters and non-adopters of an interactive electronic shopping medium," International Journal of Retail & Distribution Management, Vol.27, No.6, pp.209-223.

[69] Ekman, P. and Friesen, W. V. (1974), "Detecting deception from the body or face," Journal of Personality and Social Psychology, Vol. 29, No. 3, p.288.

[70] Eroglu, S. A., Machleit, K. A. and Davis, L. M. (2003), "Empirical testing of a model of online store atmospherics and shopper responses," Psychology & Marketing, Vol.20, No.2, pp.139-150.

[71] Fabrigar, L. R., Wegener, D. T. MacCallum, R. (1999), "Evaluating the use of exploratory factor analysis in psychological research," Psychological

Methods,Vol.4,No.3,p. 272.

[72] Finch,J. F. and West,S. G.(1997),"The investigation of personality structure:Statistical models," Journal of Research in Personality,Vol.31,No.4, pp.439–485.

[73] Fiore,A. M.,Jin,H. J. and Kim,J.(2005),"For fun and profit:Hedonic value from image interactivity and responses towards an online store," Psychology & Marketing,Vol.22,No.8,pp.669–694.

[74] Fishbein,M. and Ajzen,I.(1977),"Belief,attitude,intention and behavior: an introduction to theory and research," Philosophy and Rhetoric,Vol.10, No.2,pp:130–132.

[75] Flaviάn,C.,Guinalíu,M. and Gurrea,R.(2006),"The role played by perceived usability,satisfaction and consumer trust on website loyalty," Information & Management,Vol.43,No.1,pp.1–14.

[76] Fogg,B. J.,Kameda,T.,Boyd,J.,Marshall,J.,Sethi,R.,Sockol,M.,and Trowbridge,T.(2002),"Stanford–Makovsky web credibility study 2002:investigating what makes websites credible today," Persuasive Technology Lab.

[77] Fornell,C. and Bookstein,F. L.(1982),"Two structural equation models: LISREL and PLS applied to consumer exit–voice theory," Journal of Marketing Research,pp.440–452.

[78] Fornell,C.(1992),"A national customer satisfaction barometer:The Swedish experience," the Journal of Marketing,pp.6–21.

[79] Fornell,C. and Bookstein,F. L.(1982),"Two structural equation models: LISREL and PLS applied to consumer exit–voice theory," Journal of Marketing Research,pp.440–452.

[80] Fornell,C. and Larcker,D. F.(1981),"Structural equation models with unobservable variables and measurement error: Algebra and statistics," Journal of Marketing Research,pp.382–388.

[81] Gaeth,G. J.(1994),"Theory and method in the study of ad and brand attitudes:Towards a systemic model," Attention,Attitude,and Affect in Response to Advertising,p.125.

[82] Garrett,J. J.(2010),Elements of User Experience,the User–centered Design for The Web and Beyond,Pearson Education.

[83] Gefen,D. and Devine,P.(2001),"Customer loyalty to an online store:the meaning of online service quality," ICIS 2001 Proceedings,p. 80.

[84] Gefen, D., Straub, D. and Boudreau, M.(2000), "Structural equation modeling and regression: Guidelines for research practice," Communications of the Association for Information Systems, Vol.4, No.1, pp.7.

[85] Geuens, M., Brengman, M. and S'Jegers, R.(2003), "Food retailing, now and in the future. A consumer perspective," Journal of Retailing and Consumer Services, Vol.10, No.4, pp.241-251.

[86] Ghazali, E., Mutum, D. and Mahbob, N. A.(2006), "Exploratory study of buying fish online: are Malaysians ready to adopt online grocery shopping? " International Journal of Electronic Marketing and Retailing, Vol.1, No.1, pp.67-82.

[87] Gorsuch, R. (1983a), Factor Analysis, 2nd, Hillsdale, NJ: Lawrence Erlbaum Associates, Inc.

[88] Grazioli, S. and Jarvenpaa, S. L.(2000), "Perils of Internet fraud: An empirical investigation of deception and trust with experienced Internet consumers," IEEE Transactions on Systems, Man, and Cybernetics-Part A: Systems and Humans, Vol.30, No.4, pp.395-410.

[89] Gremler, D.(1995), "The effect of satisfaction, switching costs, and interpersonal bonds on service loyalty," Arizona State University Tempe, AZ.

[90] Griffith, D. A., Krampf, R. F. and Palmer, J. W.(2001), "The role of interface in electronic commerce: Consumer involvement with print versus on-line catalogs," International Journal of Electronic Commerce, Vol.5, No.4, pp.135-153.

[91] Guido, G., Capestro, M. and Peluso, A. M.(2007), "Experimental analysis of consumer stimulation and motivational states in shopping experiences," International Journal of Market Research, Vol.9, No.3, pp.365.

[92] Hair, J. F., Anderson, R. E.Tatham, R.(1998a), Multivariate Data Analysis, Upper Saddle River, NJ: Prentise-Hall.

[93] Hair, J. F., Ringle, C. M. and Sarstedt, M.(2013), "Editorial-partial least squares structural equation modeling: Rigorous applications, better results and higher acceptance," Long Range Planning, Vol. 46, No. 1-2, pp.1-12.

[94] Halzack, S.(2015), "The new shopping behavior that is creating big challenges for the retail industry," The Washington Post.

[95] Hansen, J., Sato, M.Ruedy, R.(2006), "Global temperature change," Proceedings of the National Academy of Sciences, Vol. 103, No. 39,

pp.14288-14293.

[96] Hansen, T. (2005), "Consumer adoption of online grocery buying: a discriminant analysis," International Journal of Retail & Distribution Management, Vol.33, No.2, pp.101-121.

[97] Hassanein, K., and Head, M. (2007), "Manipulating perceived social presence through the web interface and its impact on attitude towards online shopping," International Journal of Human-Computer Studies, Vol. 65, No.8, pp.689-708.

[98] Henseler, J., Ringle, C. M. and Sinkovics, R. R. (2009), "The use of partial least squares path modeling in international marketing," Advances in International Marketing, Vol.20, No.1, pp.277-319.

[99] Hevner, R., Daza, R.Englund, C. (2004). "Postnatal shifts of interneuron position in the neocortex of normal and reeler mice: evidence for inward radial migration," Neuroscience, Vol.124, No.3, pp.605-618.

[100] Hinkin, T. R. (1995), "A review of scale development practices in the study of organizations," Journal of Management, Vol. 21, No. 5, pp.967-988.

[101] Hirschman, E. C. (1986), "The effect of verbal and pictorial advertising stimuli on aesthetic, utilitarian and familiarity perceptions," Journal of Advertising, Vol.15, No.2, pp.27-34.

[102] Hoch, S. J. and Deighton, J. (1989), "Managing what consumers learn from experience," The Journal of Marketing, pp.1-20.

[103] Hoffman, D. L. and Novak, T. P. (1996), "Marketing in hypermedia computer-mediated environments: Conceptual foundations," The Journal of Marketing, pp.50-68.

[104] Hohlbein, D. J. and Mintel, T. E. (1999), "Toothbrush having contouring multi-component head with peel-resistant joint and limited flexibility," Google Patents.

[105] Holbrook, M. B. (1994), "The nature of customer value: an axiology of services in the consumption experience," Service Quality: New Directions in Theory and Practice, Vol.21, pp.21-71.

[106] Holbrook, M. B. and Batra, R. (1987), "Assessing the role of emotions as mediators of consumer responses to advertising", Journal of Consumer Research, pp.404-420.

[107] Hong, W., Thong, J. Y. L. and Tam, K. Y. (2004), "Does animation attract

online users' attention? The effects of flash on information search performance and perceptions," Information Systems Research, Vol.15, No.1, pp.60-86.

[107] Hui, T. K. and Wan, D.(2009), "Who are the online grocers?," The Service Industries Journal, Vol.29, No.11, pp.1479-1489.

[108] Hulland, J.(1999), "Use of partial least squares(PLS)in strategic management research: A review of four recent studies," Strategic Management Journal, Vol.20, No.2, pp. 195-204.

[109] Hwang, G. J. and Tsai, C. C.(2011), "Research trends in mobile and ubiquitous learning: A review of publications in selected journals from 2001 to 2010," British Journal of Educational Technology, Vol.42, No.4, pp.65-70.

[110] Jarvenpaa, S. L. and Dickson, G. W.(1988), "Graphics and managerial decision making: Research-based guidelines," Communications of the ACM, Vol.31, No.6, pp.764-774.

[111] Jarvenpaa, S. L. and Todd, P. A.(1996), "Consumer reactions to electronic shopping on the World Wide Web," International Journal of Electronic Commerce, Vol.1, No.2, pp.59-88.

[112] Jarvenpaa, S. L., Tractinsky, N. and Vitalec, M.(2000), "Consumer trust in an Internet store," Information Technology and Management, Vol. 1, pp.45-71.

[113] Jayawardhena, C. (2004), "Measurement of service quality in internet banking: the development of an instrument," Journal of Marketing Management, Vol.20, No.1-2, pp.185-207.

[114] Jennings, N. R.(2000), "On agent-based software engineering," Artificial intelligence, Vol.117, No.2, pp.277-296.

[115] Jiang, Z. and Benbasat, I.(2005), "Virtual product experience: Effects of visual and functional control of products on perceived diagnosticity and flow in electronic shopping," Journal of Management Information Systems, Vol.21, No.3, pp.111-147.

[116] Jiang, Z. and Benbasat, I.(2007), "Research note-investigating the influence of the functional mechanisms of online product presentations," Information Systems Research, Vol.18, No.4, pp.454-470.

[117] Jiang, Z. and Benbasat, I.(2007a), "The effects of presentation formats and task complexity on online consumers' product understanding", MIS Quarterly, pp.475-500.

[118] John D. Wells, J. S. V., Traci J. Hess(2011), "What signal are you sending? How website quality influences perceptions of product quality and purchase ", MIS Quarterly, Vol.35, No.2, pp.373-396.

[119] Kahai, S. S. and Cooper, R. B.(2003), "Exploring the core concepts of media richness theory: The impact of cue multiplicity and feedback immediacy on Decision Quality", Journal of Management Information Systems, Vol.20, No.1, pp.263-299.

[120] Kai H. Lim, L. B., Lwarence M. Ward(2000), "The role of multimedia in changing first impression bias", Information Systems Research, Vol.11, No.2, pp.115-136.

[121] Kamis, A., Koufaris, M. and Stern, T.(2008), "Using an attribute-based decision support system for user-customized products online: an experimental investigation," MIS Quarterly, pp.159-177.

[122] Kamp, E. and MacInnis, D. J.(1995), "Characteristics of portrayed emotions in commercials: When does what is shown in ads affect viewers? " Journal of Advertising Research, Vol.35, No.6, pp.19-29.

[123] Karimov, F. P. and Brengman, M.(2013), "Adoption of social media by online retailers: Assessment of current practices," Modern Entrepreneurship and E-Business Innovations, pp.41.

[124] Karimov, F. P., Brengman, M. and Van Hove, L.(2011), "The effect of website design dimensions on initial trust: a synthesis of the empirical literature," Journal of Electronic Commerce Research, Vol.12, No.4, pp.272.

[125] Karvonen, K. Year.(2000), "The beauty of simplicity," Proceedings on the 2000 Conference on Universal Usability, ACM2000, pp.85-90.

[126] Kathy, H., Robert, E. and Ehrenberg Andrew, S.(1996), "Buying more and buying longer: concepts and applications of consumer loyalty," London: London Business School.

[127] Kempf, D. S.(1999), "Attitude formation from product trial: Distinct roles of cognition and affect for hedonic and functional products," Psychology & Marketing, Vol.16, No.1, pp.35-50.

[128] Kempf, D. S. and Smith, R. E.(1998), "Consumer processing of product trial and the influence of prior advertising: A structural modeling approach," Journal of Marketing Research, pp.325-338.

[129] Keppel, G.(1991), Design and Analysis: A Researcher's Handbook, Prentice-Hall, Inc.

[130] Kerlinger,F. and Lee,H.(2000),"Survey research," Foundations of Behavioral Research,pp.599-619.

[131] Kim,J. and Moon,J. Y.(1998),"Designing towards emotional usability in customer interfaces—trustworthiness of cyber-banking system interfaces," Interacting with Computers,Vol.10,No.1,pp.1-29.

[132] Kim,M. and Lennon,S. J.(2000),"Television shopping for apparel in the United States: effects of perceived amount of information on perceived risks and purchase intentions", Family and Consumer Sciences Research Journal,Vol.28,No.3,pp.301-331.

[133] Kim,S. and Stoel,L.(2004),"Apparel retailers: website quality dimensions and satisfaction," Journal of Retailing and Consumer Services, Vol. 11, No.2,pp.109-117.

[134] Kim,W. G. and Moon,Y. J.(2009),"Customers' cognitive,emotional,and actionable response to the services cape: A test of the moderating effect of the restaurant type," International Journal of Hospitality Management , Vol.28,No.1,pp .144-156.

[135] Kirk,T.,Levy,A. Y.Sagiv,Y.(1995),"The information manifold," Proceedings of the AAAI1995,pp.85-91.

[136] Kline,R.(2010),Principles and Practice of Structural Equation Modeling, 3rd edn Guilford Press,New York. USA.

[137] Knutson,J. F.(1998),"The Effect of the user interface design on adoption of new technology," Georgia Institute of Technology.

[138] Koufaris,M.(2002),"Applying the technology acceptance model and flow theory to online consumer behavior," Information Systems Research, Vol.13,No.2,pp.205-223.

[140] Kurnia,S.(2008),"Exploring e-commerce readiness in China: The case of the grocery industry," Hawaii International Conference on System Sciences,Proceedings of the 41st Annual,IEEE2008,pp.413-413.

[141] Lam,S. Y.,Shankar,V.Erramilli,M. K.(2004),"Customer value,satisfaction,loyalty,and switching costs: an illustration from a business-to-business service context," Journal of the Academy of Marketing Science, Vol.32,No.3,pp.293-311.

[142] Larkin,J. H. and Simon,H. A.(1987),"Why a diagram is (sometimes) worth ten thousand words," Cognitive Science,Vol.11,No.1,pp.65-100.

[143] Laroche,M.,Yang,Z.McDougall,G. H.(2005),"Internet versus bricks-

and - mortar retailers: An investigation into intangibility and its consequences," Journal of Retailing, Vol.81, No.4, pp.251-267.

[144] Lavie, T., Oron-Gilad, T. and Meyer, J.(2011). "Aesthetics and usability of in-vehicle navigation displays," International Journal of Human-Computer Studies, Vol.69, No.1, pp.80-99.

[145] Lee, M. K., Cheung, C. M. and Chen, Z.(2005), "Acceptance of Internet-based learning medium: the role of extrinsic and intrinsic motivation," Information & Management, Vol.42, No.8, pp.1095-1104.

[146] Lee, U. K.(2015), "The effect of information deception in price comparison site on the consumer reactions: An empirical verification," International Journal of Distributed Sensor Networks.

[147] Lee, W. and Benbasat, I.(2003), "Designing an electronic commerce interface: attention and product memory as elicited by web design," Electronic Commerce Research and Applications, Vol.2, No.3, pp.240-253.

[148] Lewis, B. R. and Mitchell, V. W.(1990), "Defining and measuring the quality of customer service," Marketing Intelligence & Planning, Vol.8, No.6, pp.11-17.

[149] Li, X., Wang, M. and Chen, Y.(2014), "The impact of product photo on online consumer purchase intention: an image-processing enabled empirical study," PACIS2014, pp.325-344.

[150] Lightner, N. J. and Eastman, C. M.(2002), "User preference for product information in remote purchase environments," Journal of Electronic Commerce Research, Vol.3, No.3, pp .174-186.

[151] Lim, H., Widdows, R. and Hooker, N. H.(2009), "Web content analysis of e-grocery retailers: a longitudinal study," International Journal of Retail & Distribution Management, Vol.37, No.10, pp.839-851.

[152] Lim, K. H.Benbasat, I. and Ward, L. M.(2000), "The role of multimedia in changing first impression bias," Information Systems Research, Vol.11, No.2, pp.115-136.

[153] Lin, Y. C.Yeh, C. H. and Wei, C. C.(2013), "How will the use of graphics affect visual aesthetics? A user-centered approach for webpage design," International Journal of Human-Computer Studies, Vol.71, No.3, pp.217-227.

[154] Lindgaard, G.(1999), "Does emotional appeal determine perceived usability of websites," Proceedings of CybErg: the Second International Cyberspace Conference on Ergonomics 1999, pp.202-211.

[155] Lindgaard, G. and Dudek, C. (2003), "What is this evasive beast we call user satisfaction?" Interacting with Computers, Vol.15, No.3, pp.429-452.

[156] Lindgaard, G., Fernandes, G. Dudek, C. (2006), "Attention web designers: You have 50 milliseconds to make a good first impression!" Behaviour & Information Technology, Vol.25, No.2, pp.115-126.

[157] Liu, C., Arnett, K. P. and Capella, L. M. (2001), "Key dimensions of web design quality as related to consumer response," The Journal of Computer Information Systems, Vol.42, No.1, pp.70.

[158] Lo, L. and Lin, S. W. (2013), "Three ways to convert browsing into impulse buying: Website streamline and decoration," active citizenship by knowledge management & innovation: Proceedings of the Management, Knowledge and Learning International Conference 2013, To Know Press 2013, pp.1177-1186.

[159] Loehlin, J. C. (1992), Genes and Environment in Personality Development, Sage Publications, Inc.

[160] Loiacono, E. T., Watson, R. T. and Goodhue, D. L. (2000), "WebQual: a website quality instrument," University of Georgia Athens, GA.

[161] Loiacono, E. T., Watson, R. T. and Goodhue, D. L. (2007), "WebQual: An instrument for consumer evaluation of websites," International Journal of Electronic Commerce, Vol.11, No.3, pp.51-87.

[162] Lou, H., Chau, P. Y. and Li, D. (2005), "Understanding individual adoption of instant messaging: An empirical investigation," Journal of the Association for Information Systems, Vol.6, No.4, pp.5.

[163] Mahon-Haft, T. A. and Dillman, D. A. (2010), "Does visual appeal matter? Effects of web survey aesthetics on survey quality," Survey Research Methods 2010, pp.43-59.

[167] Malhotra, N. K., Agarwal, J. and Peterson, M. (1996), "Methodological issues in cross-cultural marketing research: A state-of-the-art review," International Marketing Review, Vol.13, No.5, pp.7-43.

[168] Malhotra, N. K., Kim, S. S. and Patil, A. (2006), "Common method variance in IS research: A comparison of alternative approaches and a reanalysis of past research," Management Science, Vol.52, No.12, pp.1865-1883.

[169] Mathwick, C., Malhotra, N. and Rigdon, E. (2001), "Experiential value: conceptualization, measurement and application in the catalog and Internet shopping environment," Journal of Retailing, Vol.77, No.1, pp.39-56.

[170] Mathwick, C. and Rigdon, E. (2004), "Play, flow, and the online search experience," Journal of Consumer Research, Vol.31, No.2, pp.324–332.

[171] Mayer, R. E., Dow, G. T. and Mayer, S. (2003), "Multimedia learning in an interactive self – explaining environment: What works in the design of agent-based micro worlds? " Journal of Educational Psychology, Vol.95, No.4, p. 806.

[172] McKinney, V., Yoon, K. and Zahedi, F. M. (2002), "The measurement of web – customer satisfaction: An expectation and disconfirmation approach," Information Systems Research, Vol.13, No.3, pp.296–315.

[173] McKinsey, J. (2016b), "Isomorphism of games and strategic equivalence," Contributions to the Theory of Games I, pp.117–130.

[174] Menon, S. and Kahn, B. (2002), "Cross-category effects of induced arousal and pleasure on the Internet shopping experience," Journal of Retailing, Vol.8, No.1, pp.31–40.

[175] Meyers-Levy, J. and Peracchio, L. A. (1992a), "Getting an angle in advertising: The effect of camera angle on product evaluations," Journal of Marketing Research, pp.454–461.

[176] Meyers-Levy, J. and Peracchio, L. A. (1992b), "Getting an angle in advertising: The effect of camera angle on product evaluations", Journal of Marketing Research, pp.454–461.

[177] Montoya-Weiss, M. M. Voss, G. B. and Grewal, D. (2003), "Determinants of online channel use and overall satisfaction with a relational, multichannel service provider," Journal of the Academy of Marketing Science, Vol.31, No.4, pp.448–458.

[178] Moon, J. W. and Kim, Y. G. (2001), "Extending the TAM for a World-Wide-Web context," Information & management, Vol.38, No.4, pp.217–230.

[179] Morganosky, M. A. and Cude, B. J. (2000), "Consumer response to online grocery shopping," International Journal of Retail & Distribution Management, Vol.28, No.1, pp.17–26.

[180] Morrison, J. and Vogel, D. (1998), "The impacts of presentation visuals on persuasion," Information & Management, Vol.33, No.3, pp.125–135.

[181] Mortimer, G., Fazal e Hasan, S. Andrews, L. (2016), "Online grocery shopping: the impact of shopping frequency on perceived risk," The International Review of Retail, Distribution and Consumer Research, pp.1–22.

[182] Nielsen 2015. http://www.nielsen.com.

[183] Nisbett, R. E. and Ross, L. (1980), Human Inference: Strategies and Shortcomings of Social Judgment, Englewood Cliffs, NJ: Prentice-Hall, p.333.

[184] Nixon, M., Nixon, M. S. and Aguado, A. S. (2012), Feature Extraction & Image Processing for Computer Vision, Academic Press.

[185] Nordfält, J., Grewal, D.Roggeveen, A. L. (2014), "Insights from in-store marketing experiments," Review of Marketing Research, Vol.11, pp.127-146.

[186] Norris, M. and Lecavalier, L. (2010), "Evaluating the use of exploratory factor analysis in developmental disability psychological research," Journal of Autism and Developmental Disorders, Vol.40, No.1, pp.8-20.

[187] Nothdurft, H. C. (1993), "Saliency effects across dimensions in visual search," Vision Research, Vol.33, No.5-6, pp .839-844.

[188] Novak, T. P., Hoffman, D. L. and Yung, Y.-F. (2000), "Measuring the customer experience in online environments: A structural modeling approach," Marketing Science, Vol.19, No.1, pp.22-42.

[189] Nunnally, J. (1978a), Psychometric Theory, New York: McGraw-Hill.

[190] Nunnally, J. (1978b), Psychometric Methods, New York: McGraw-Hill.

[191] Nunnally, J. C. (1978c), Psychometric Theory (Second Edition), New York, McGraw-Hill.

[192] Nunnally, J. C. and Bernstein, I. (1994), "The assessment of reliability," Psychometric Theory, Vol.3, No.1, pp.248-292.

[193] O'keefe, R. M. and McEachern, T. (1998), "Web-based customer decision support systems," Communications of the ACM, Vol. 41, No. 3, pp.71-78.

[194] Oliver, R. L. (1999), "Whence consumer loyalty? " the Journal of Marketing, pp .33-44.

[195] Ou, C. X. and Sia, C. L. (2010), "Consumer trust and distrust: An issue of website design," International Journal of Human-Computer Studies, Vol.68, No.12, pp.913-934.

[196] Parasuraman, A., Berry, L. L. and Zeithaml, V. A. (1991), "Refinement and reassessment of the SERVQUAL scale," Journal of Retailing, Vol. 67, No.4, p. 420.

[197] Parasuraman, A., Zeithaml, V. A. and Berry, L. (1988), "Servqual," Journal of Retailing, Vol.64, No.1, pp.12-40.

[198] Parboteeah, D. V., Valacich, J. S. and Wells, J. D. (2009), "The influence of website characteristics on a consumer's urge to buy impulsively," Infor-

mation Systems Research, Vol.20, No.1, pp.60–78.

[199] Park, J., Lennon, S. J. and Stoel, L. (2005), "On-line product presentation: Effects on mood, perceived risk, and purchase intention," Psychology and Marketing, Vol.22, No.9, pp.695–719.

[200] Park, J., Stoel, L. and Lennon, S. J. (2008), "Cognitive, affective and conative responses to visual simulation: The effects of rotation in online product presentation", Journal of Consumer Behavior, Vol.7, No.1, pp.72–87.

[201] Pavlou, P. (2001), "Integrating trust in electronic commerce with the technology acceptance model: model development and validation," AMCIS 2001 Proceedings, p. 159.

[202] Peracchio, L. A. and Meyers-Levy, J. (1994), "How ambiguous cropped objects in ad photos can affect product evaluations," Journal of Consumer Research, pp.190–204.

[203] Perea y Monsuwé, T. Dellaert, B. G. and De Ruyter, K. (2004), "What drives consumers to shop online? A literature review," International Journal of Service Industry Management, Vol.15, No.1, pp.102–121.

[204] Pereverzeva, M. and Teller, D. Y. (2009), "Simultaneous color contrast in 4-month-old infants," Perception, Vol.38, No.1, pp.30–43.

[205] Raijas, A. (2002), "The consumer benefits and problems in the electronic grocery store," Journal of Retailing and Consumer Services, Vol.9, No.2, pp.107–113.

[206] Ramus, K. and Asger Nielsen, N. (2005), "Online grocery retailing: what do consumers think?," Internet Research, Vol.15, No.3, pp.335–352.

[207] Raney, A. A., Arpan, L. M. Pashupati, K. (2003), "At the movies, on the web: An investigation of the effects of entertaining and interactive web content on site and brand evaluations," Journal of Interactive Marketing, Vol.14, No.7, pp.38–53.

[208] Rankin, J. (2013), "Online Grocery Shopping: When will Morrison come in from the Cold? " The Guardian, Vol.8.

[209] Reber, R. and Schwarz, N. (1999), "Effects of perceptual fluency on judgments of truth," Consciousness and Cognition, Vol.8, No.3, pp.338–342.

[210] Reeves, B. and Nass, C. (1996), How People Treat Computers, Television, and New Media Like Real People and Places," CSLI Publications and Cambridge University Press.

[211] Reichheld, F. F. and Sasser, W. E. (1990), "Zero defections: quality comes

to services". Harvard Business Review, Vol.68, No.5, pp: 105-112.

[212] Richards, B. (1996), "Technology (a special report): let the buying begin. How big a bite? " The Wall Street Journal, Vol.17, p. 10.

[213] Riegelsberger, J. (2002), "The effect of facial cues on trust in e-commerce systems," Proceedings 2002.

[214] Riegelsberger, J. Sasse, M. A. and McCarthy, J. D. (2003), "Shiny happy people building trust? Photos on e-commerce website and consumer trust," Proceedings of the SIGCHI Conference on Human Factors in Computing Systems, ACM2003, pp.121-128.

[215] Ringle, C., Wende, S. and Will, A. (2005), SmartPLS 2.0 (Beta), Hamburg.

[216] Robinson, H., Dall'Olmo, R. Rettie, R. (2007), "The role of situational variables in online grocery shopping in the UK, " The Marketing Review, Vol.7, No.1, pp.89-106.

[217] Rosen, D. E. and Purinton, E. (2004), "Website design: Viewing the web as a cognitive landscape," Journal of Business Research, Vol. 57, No. 7, pp.787-794.

[218] Russell-Bennett, R., J. R and Coote, L. V. (2007), "Involvement, satisfaction, and brand loyalty in a small business services setting, " Journal of Business Research, Vol.60, No.12, pp.1253-1260.

[219] Salinas, C. (2002), "Technical rhetoricians and the art of configuring images," Technical Communication Quarterly, Vol.11, No.2, pp.165-183.

[220] Sarah Sungsook Song, M. K. (2012), "Does more mean better an examination of visual product presentation in e-retailing", Journal of Electronic Commerce Research, Vol.13, No.4, pp.345-355.

[221] Scapin, D. L. and Bastien, J. C. (1997), "Ergonomic criteria for evaluating the ergonomic quality of interactive systems, " Behaviour & Information Technology, Vol.16, No.5, pp.220-231.

[222] Scott, L. M. (1994), "Images in advertising: The need for a theory of visual rhetoric," Journal of Consumers Research, Vol.21, No.2, pp.252-273.

[223] Shavitt, S., Lowrey, P. and Haefner, J. (1998), "Public attitudes towards advertising: More favorable than you might think", Journal of Advertising Research, Vlo.38, No.4, pp.7-22.

[224] Shih, J. C. (2004), "Cesarean scar pregnancy: diagnosis with three-dimensional (3D) ultrasound and 3D power Doppler," Ultrasound in Obstetrics & Gynecology, Vol.23, No.3, pp.306-307.

[225] ShopperVista 2013. "Online grocery shoppers," Factsheet. IGD.

[226] Singleton Jr, R. and Straits, B. (1999), Approaches to Social Research, New York: Oxford Univ Press.

[227] Son, J. Y. and Kim, S. S. (2008), "Internet users' information privacy-protective responses: A taxonomy and a nomological model," MIS Quarterly, pp.503–529.

[228] Sosik J J, Kahai S S, Piovoso M. J. (2009), "Silver bullet or voodoo statistics? A primer for using the partial least squares data analytic technique in group and organization research," Group & Organization Management, Vol.34, No.1, pp.5–36.

[229] Suh, K.-S. and Lee, Y. E. (2005), "The effects of virtual reality on consumer learning: an empirical investigation," MIS Quarterly, pp.673–697.

[230] Sun, H. and Zhang, P. (2008), "An exploration of affect factors and their role in user technology acceptance: Mediation and causality," Journal of the American Society for Information Science and Technology, Vol. 59, No.8, pp.1252–1263.

[231] Sundar, S. and Kalyanaraman, S. (2004), "Arousal, memory, and impression-formation effects of animation speed in web advertising," Journal of Advertising, Vol.33, No.1, pp.7–17.

[232] Sussman, S. W. and Siegal, W. S. (2003), "Informational influence in organizations: An integrated approach to knowledge adoption," Information Systems Research, Vol.14, No.1, pp.47–65.

[233] Swinyard, W. R. (1993). "The effects of mood, involvement, and quality of store experience on shopping intentions", Journal of Consumer Research, pp.271–280.

[234] Szymanski, D. M. and Hise, R. T. (2000), "E-satisfaction: an initial examination," Journal of Retailing, Vol.76, No.3, pp.309–322.

[235] Tan, F. B., Tung, L. L. and Xu, Y. (2009), "A study of Web-designers' criteria for effective business-to-consumer (B2C) Website using the repertory grid technique," Journal of Electronic Commerce Research, Vol. 10, No.3, p. 155.

[236] Tarafdar, M. and Zhang, J. (2005), "Analysis of critical website characteristics: A cross-category study of successful website," Journal of Computer Information Systems, Vol.46, No.2, pp.14–24.

[237] Taylor, S. E. and Thompson, S. C. (1982), "Stalking the elusive 'vividness'

effect," Psychological Review, Vol.89, No.2, p. 155.

[238] Terwogt, M. M. and Hoeksma, J. B. (1995), "Colors and emotions: Preferences and combinations," The Journal of General Psychology, Vol. 122, No.1, pp.5-17.

[239] Tharenou, P., Donohue, R. and Cooper, B. (2007), "Management research methods," Cambridge University Press New York, NY.

[240] Tseng, S. and Fogg, B. (1999), "Credibility and computing technology," Communications of the ACM, Vol.42, No.5, pp.39-44.

[241] Turley, L. W. and Milliman, R. E. (2000), "Atmospheric effects on shopping behavior: a review of the experimental evidence," Journal of Business Research, Vol.49, No.2, pp.193-211.

[242] Van der Heijden, H. (2003), "Factors influencing the usage of website: the case of a generic portal in the Netherlands," Information & Management, Vol.40, No.6, pp.541-549.

[243] Van der Heijden, H. (2004), "User acceptance of hedonic information systems," MIS Quarterly, pp.695-704.

[244] Van der Heijden, H. Verhagen, T. and Creemers, M. (2003), "Understanding online purchase intentions: contributions from technology and trust perspectives," European Journal of Information Systems, Vol. 12, No. 1, pp.41-48.

[245] Verhagen, T., Vonkeman, C. Feldberg, F. (2014), "Present it like it is here: Creating local presence to improve online product experiences," Computers in Human Behavior, Vol.39, pp.270-280.

[246] Verhoef, P. C. and Langerak, F. (2001), "Possible determinants of consumers' adoption of electronic grocery shopping in the Netherlands," Journal of Retailing and Consumer Services, Vol.8, No.5, pp.275-285.

[247] Vieira, V. A. (2010), "Visual aesthetics in store environment and its moderating role on consumer intention," Journal of Consumer Behaviour, Vol.9, No.5, pp.364-380.

[248] Vinzi, V. E., Trinchera, L. and Amato, S. (2010), "PLS path modeling: from foundations to recent developments and open issues for model assessment and improvement," Springer Berlin Heidelberg, pp.47-82.

[249] Webster, J. and Ho, H. (1997), "Audience engagement in multimedia presentations," ACM SIGMIS Database, Vol.28, No.2, pp.63-77.

[250] Wells, J. D., Parboteeah, V. and Valacich, J. S. (2011a), "Online impulse

buying: understanding the interplay between consumer impulsiveness and website quality," Journal of the Association for Information Systems, Vol.12. No.1, p. 32.

[251] Wells, J. D., Valacich, J. S. and Hess, T. J. (2011b), "What signals are you sending? How Website quality influences perceptions of product quality and purchase intentions," MIS Quarterly, Vol.35, No.2, pp.373–396.

[252] Wold, H. (1985), Systems Analysis by Partial Least Squares, . Dordrecht, the Netherlands: Martinus Nijhoff, pp.221–252.

[253] Wong, K. K. (2011), "Book review: handbook of partial least squares: concepts, methods and applications," International Journal of Business Science & Applied Management. Vol.6, No.2, pp.52–54.

[254] Wong, K. K. (2010), "Handling small survey sample size and skewed dataset with partial least square path modeling," Marketing Research and Intelligence Association, pp.20–23.

[255] Wu, C.S., Cheng, F. F. and Yen, D. C. (2008), "The atmospheric factors of online storefront environment design: An empirical experiment in Taiwan," Information & Management, Vol.45, No.7, pp .493–498.

[256] Xiao, B. (2010), Product-related Deceptive Information Practices in B2C E-commerce: Formation, Outcomes, and Detection, University of British Columbia.

[257] Xu, J. D., Benbasat, I. and Cenfetelli, R. T. (2013), "Integrating service quality with system and information quality: an empirical test in the e-service context," MIS Quarterly, Vol.37, No.3, pp.777–794.

[258] Yoo, J. and Kim, M. (2012), "Online product presentation: the effect of product coordination and a model's face," Journal of Research in Interactive Marketing, Vol.6, No.1, pp.59–72.

[259] Yoo, J. and Kim, M. (2014), "The effects of online product presentation on consumer responses: A mental imagery perspective," Journal of Business Research, Vol.67, No.11, pp.2464–2472.

[260] Zeithaml, V. A. Berry, L. L. and Parasuraman, A. (1996), "The behavioral consequences of service quality," the Journal of Marketing, pp.31–46.

[261] Zhang, P., Benbasat, I. Carey, J. (2002), "AMCIS 2002 panels and workshops I: Human – computer interaction research in the MIS discipline," Communications of the Association for Information Systems, Vol.9, No.1, p.20.

[262] Zhang, P. and Dillon, A. (2003), "HCI and MIS: shared concerns," International Journal of Human-Computer Studies, Vol.59, No.4, pp.397-402.

[263] Zhang, P., Von Dran, G. M. Blake, P. (2001), "Important design features in different Website domains: An empirical study of user perceptions," E-Service, Vol.1, No.1, pp.77-91.

[264] Zhenhui. Jiang, I. B. (2007b), "Investigating the influence of the functional mechanisms of online product presentations," Information Systems Research, Vol.18, No.4, pp.454-470.

APPENDIX A

Examples of the Webpage for the Online Grocery Store

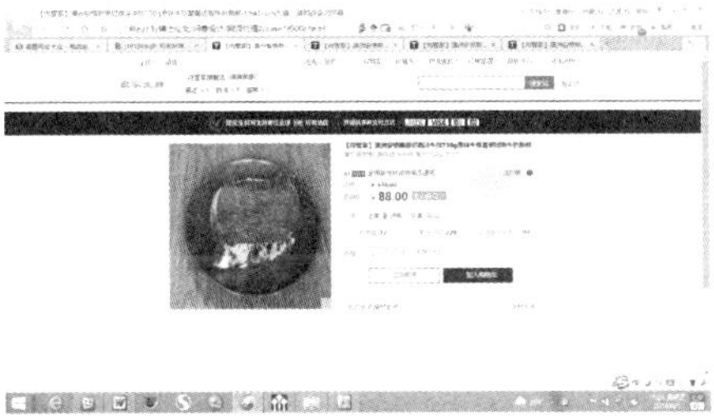

Case 1. Reflective Surface (0), Contrast Color Usage (0), Complementary Goods Layout (0)

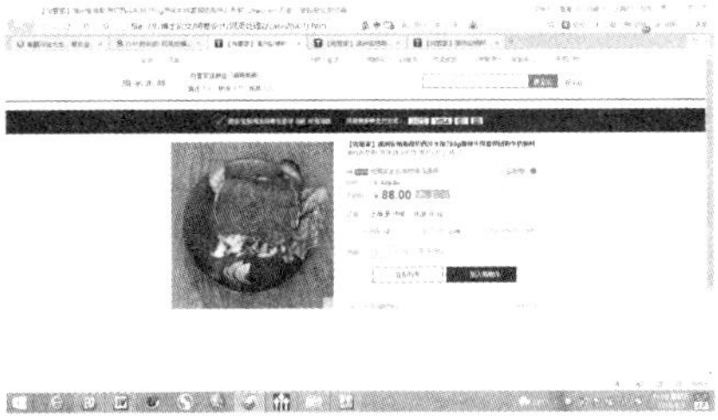

Case 2. Reflective Surface (0), Contrast Color Usage (0), Complementary Goods Layout (1)

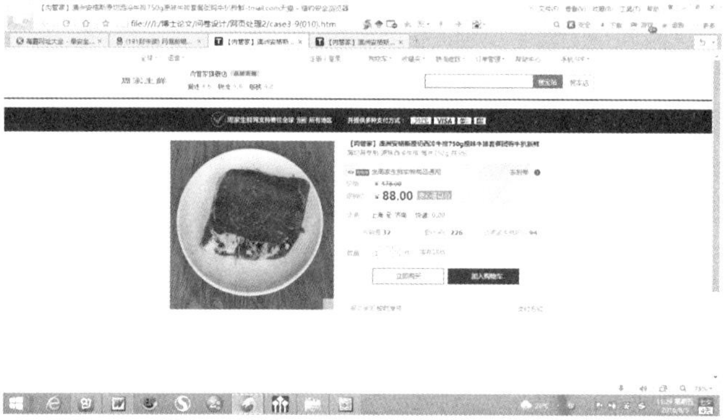

Case 3. Reflective Surface（0），Contrast Color Usage（1），Complementary Goods Layout（0）

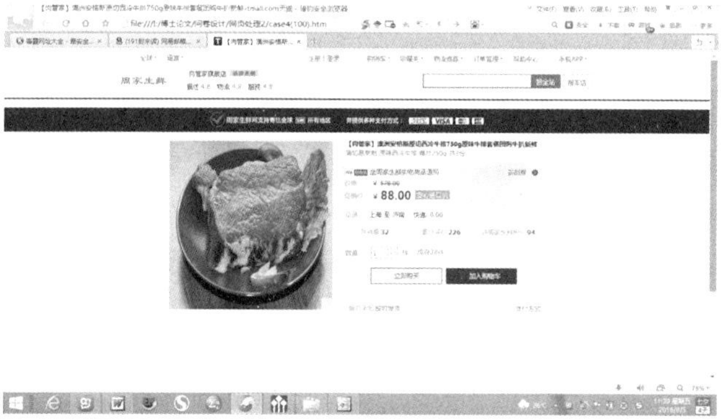

Case 4. Reflective Surface（1），Contrast Color Usage（0），Complementary Goods Layout（1）

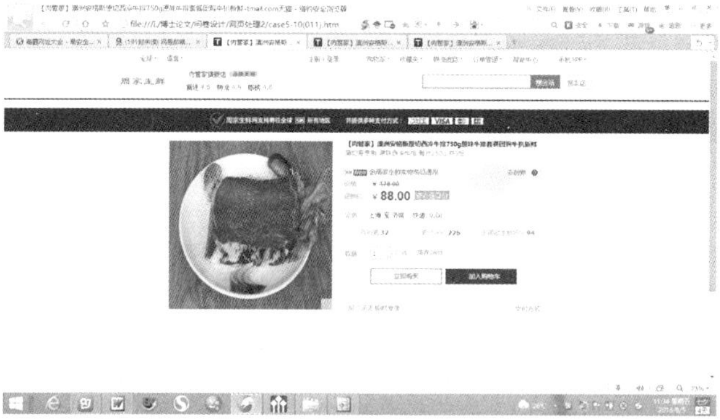

Case 5. Reflective Surface（0），Contrast Color Usage（1），Complementary Goods Layout（1）

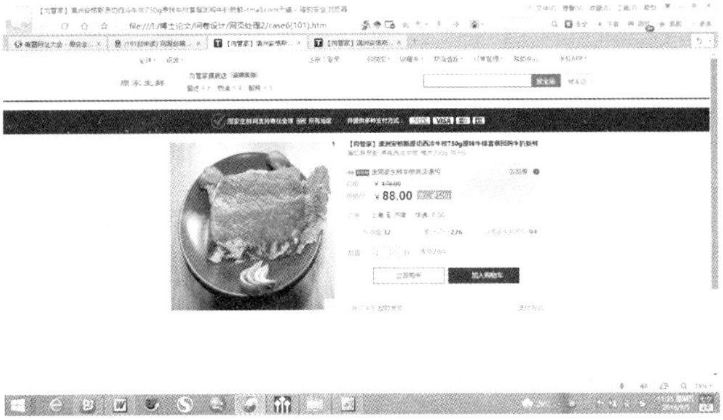

Case 6. Reflective Surface (1), Contrast Color Usage (0), Complementary Goods Layout (1)

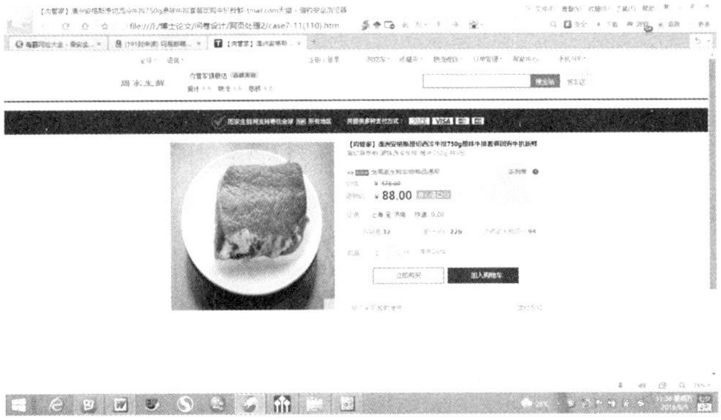

Case 7. Reflective Surface (1), Contrast Color Usage (1), Complementary Goods Layout (0)

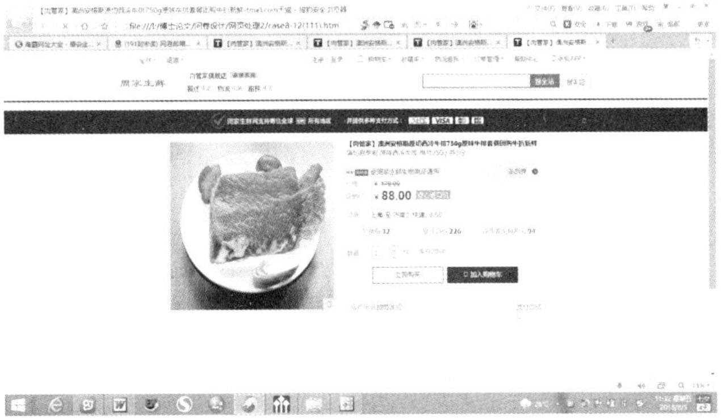

Case 8. Reflective Surface (1), Contrast Color Usage (1), Complementary Goods Layout (1)

APPENDIX B

Items for Measuring Constructs

Constructs	Measurement items	Sources
Visual Appeal (VA)	(VA1) The online grocery photo is visually pleasing. (VA2) The online grocery photo displays a visually pleasing design. (VA3) The online grocery photo is aesthetically appealing. (VA4) The way the grocery photo displays its products is attractive. (VA5) I like the way the online grocery's photo looks.	Loiacono et al. (2007) Mathwick et al. (2001)
Shopping Enjoyment (SE)	(SE1) I find my experience with this website interesting. (SE2) I find my experience with this website enjoyable. (SE3) I find my experience with this website is exciting. (SE4) I find my experience with this website fun.	Koufaris (2001—2002)
Attitude Towards Product (ATP)	(ATP1) The product that I've just examined is good. (ATP2) I have formed a favorable impression towards the product that I' ve just examined. (ATP3) I like the product that I've just examined.	Jiang and Benbasat (2007b)
Attitude Towards Store (ATS)	(ATS1) I like shopping on this online grocery store. (ATS2) Shopping on this online grocery store is a good idea. (ATS3) Shopping on this online grocery store is appealing. (ATS4) I think this grocery store is favorable.	Jiang and Benbasat (2007b) Grazioli and Jarvenpaa (2000)

续表

Constructs	Measurement items	Sources
Purchase Intention (PI)	(PI1) It is likely that I will buy this grocery item. (PI2) I will definitely try this grocery item. (PI3) I have strong intentions to purchase the grocery item. (PI4) The probability that I would consider buying at this grocery from website is high. (PI4) If I were to buy grocery item, I would consider buying it from this store.	Coyle and Thorson (2001) Jiang and Benbasat (2007b)
Privacy Concern (PC)	(PC1) I am concerned that the information I submit on the Internet could be misused. (PC2) I am concerned that a person can find private information about me on the Internet. (PC3) I am concerned about providing personal information on the Internet, because I do not know what others might do with it. (PC4) I am concerned about providing personal information on the Internet, because it could be used in a way I did not foresee.	Son and Kim (2008)
Issue Involvement (IV)	(IV1) I am much involved in the topic of photo decoration cues in online grocery store. (IV2) Many of the issues discussed about photo decoration cues would be on my mind lately.	Sussman and Siegal (2003)
Expertise (EP)	(EP1) I am much informed on the subject matter of the photo decoration cues in online grocery store. (EP2) I am an expert on the topic of photo decoration cues in online grocery store.	Sussman and Siegal (2003)
Information Deception Detection (IDD)	Overall, I believe that the given product presented in this online grocery store are ... (IDD1) Misleading. (IDD2) Deceptive. (IDD3) Biased.	Xiao (2010)

Notes: 'VA' stands for visual appeal. 'SE' stands for shopping enjoyment. 'ATP' stands for attitude towards product. 'ATS' stands for attitude towards store. 'PI' stands for purchase intention. 'PC' stands for privacy concern. 'IV' stands for issue involvement. 'EP' stands for expertise. 'IDD' stands for information deception detection.

APPENDIX C

变量	描述	来源
视觉吸引	这张图片令人赏心悦目。 这张图片的设计令人赏心悦目。 从审美角度看，这张图片很吸引人。 这张图片展示产品的方式很吸引人。 我喜欢这张图片展示出来的样子	Loiacono et al.（2007） Mathwick et al.（2001）
购物乐趣	体验这个网站很有趣。 体验这个网站令人愉快。 体验这个网站令人兴奋。 体验这个网站令人感觉好玩	Koufaris（2001—2002）
对产品的态度	刚才看过的牛排不错。 刚才看过的牛排给我留下了好的印象。 我喜欢刚才看到的牛排	Jiang and Benbasat（2007b）
对网站的态度	我喜欢在这个生鲜网站上购物。 在这个生鲜网站上购物是个好主意。 在这个生鲜网站上购物很吸引人。 我认为这个生鲜网站受人喜爱	Jiang and Benbasat（2007b） Grazioli and Jarvenpaa（2000）

续表

变量	描述	来源
购买牛排的意愿	我很可能会买这款牛排。 我肯定要买这款牛排。 我买这款牛排的意愿很强烈。 我很可能会考虑购买这款牛排。 如果要买牛排，我会考虑在这家网站上购买	Coyle and Thorson (2001) Jiang and Benbasat (2007b)
对网站的忠诚度	每当我要购买牛排的时候，我就尽可能到这个网站。 每当我要购买牛排的时候，我首选这个网站。 我喜欢用这个生鲜网站。 对我来说，这个网站是网上买生鲜的最佳选择。 我想这是我最喜欢的牛排购物网站。 下次我要买牛排的时候，我愿意使用这个网站	Zeithamlet al. (1996) Coyle and Thorson (2001) Jiang and Benbasat (2007b)
对个人隐私的看法	我担心提交到网上的个人信息可能会被滥用。 我担心有人能在网上发现我的个人隐私。 我担心提交到网上的个人信息，会被别人利用。 我担心提交到网上的个人信息，以某种不可预料的方式被利用	Son and Kim (2008)
对主题的了解程度	我对生鲜食品图片装饰设计这个相关的主题接触颇多。 近来，与图片装饰设计相关的很多问题，我都有所思考	Sussman and Siegal（2003）
专业知识了解程度	关于生鲜食品图片装饰设计的这个话题，我了解非常多。 对于生鲜食品图片装饰设计相关的话题，我有专业的了解	Sussman and Siegal（2003）
信息欺骗检测意见	误导 欺骗 有偏差	Xiao（2010）

APPENDIX D

Online Survey 1~2

Dear Friends,

Thank you for taking part in our online survey.

We would like you to answer three questions about your opinions of online grocery photo. All answers will remain anonymous and we want to emphasize that there are no right or wrong answers. All we are interested in are your opinions.

Thanks for your support!

Online Survey 1

Two pieces of beef are chosen as the target grocery products presented as follow. Please rate the response that best reflects your opinion about the effect of photo's reflective surface, contrast color usage and complementary goods layout.

Questions	Yes	No
1. Do you think there are some reflective effects on the surface of beef in the above photo?	○	○
2. Do you think there is color contrast effect between the color of beef and plate in the above photo?	○	○

Questions	Yes	No
3. Do you think there are some complementary goods of the beefsteak in the above photo?	○	○

Online Survey 2

Two pieces of beef are chosen as the target grocery products presented as follow. Please rate the response that best reflects your opinion about the effect of photo's reflective surface, contrast color usage and complementary goods layout.

Questions	Yes	No
1. Do think there are some reflective effects on the surface of beef in the above photo?	○	○
2. Do you think there is color contrast effect between the color of beef and plate in the above photo?	○	○

Questions	Yes	No
3. Do you think there are some complementary goods of the beefsteak in the above photo?	○	○

APPENDIX E

生鲜食品图片调查问卷（1）

亲爱的朋友：

您好！欢迎您参加本次调查。我们希望能从您那里了解一些关于生鲜食品图片方面的线索。下面是两张牛排图片，请对图片中是否存在反射光效果，对比色的使用和食材布局做出您的评价（单选题）。

题目	是	否
1.在上图中，您认为牛排表面有反射光效果吗？	○	○
2.在上图中，您认为盘子的颜色与牛排的颜色之间存在对比色吗？	○	○

题目	是	否
3.在上图中，您认为洋葱、生菜和香菇等配菜与牛排之间存在着一种布局关系吗？	○	○

生鲜食品图片调查问卷（2）

亲爱的朋友：

您好！欢迎您参加本次调查。我们希望能从您那里了解一些关于生鲜食品图片方面的线索。下面是两张牛排图片，请对图片中是否存在反射光效果，对比色的使用和食材布局做出您的评价（单选题）。

题目	是	否
1.在上图中，您认为牛排表面有反射光效果吗？	○	○
2.在上图中，您认为盘子的颜色与牛排的颜色之间存在对比色吗？	○	○

题目	是	否
3.在上图中，您认为洋葱、生菜和香菇等配菜与牛排之间存在着一种布局关系吗？	○	○

APPENDIX F

Online Grocery Shopping Survey Questionnaire

Dear Friends,

Thank you for taking part in our online survey.

We would like you to answer some questions about your opinions of online grocery shopping. It should take about 15–20 minutes.

Imagine we develop a virtual online grocery store (a website) that you will shop grocery on this website. A piece of beefsteak sold on the above website is chosen as the target grocery product presented as follow. Please look over the photo carefully and familiarize yourself with it prior to beginning the survey. How much do you prefer each decoration cue associated with this photo?

Please rate the response that best reflects your opinion of the photo and virtual website you just examined. All answers will remain anonymous and we want to emphasize that there are no right or wrong answers. All we are interested in are your opinions (From level 1 to 7, 1 is strongly disagree, 7 is strongly agree).

1. To what extent do you think that the photo decoration cues affect your visual appeal?

Questions	Responses						
	Strongly disagree				Strongly agree		
The online grocery photo is visually pleasing	1	2	3	4	5	6	7
The online grocery photo displays visually pleasing design	1	2	3	4	5	6	7
The online grocery photo is aesthetically appealing.	1	2	3	4	5	6	7
The way the grocery photo displays its products is attractive	1	2	3	4	5	6	7
I like the way the online grocery' s photo looks	1	2	3	4	5	6	7

2. To what extent do you think that the photo decoration cues affect your shopping enjoyment?

Questions	Responses						
	Strongly disagree				Strongly agree		
I find my experience with this website interesting	1	2	3	4	5	6	7
I find my experience with this website enjoyable	1	2	3	4	5	6	7
I find my experience with this website exciting	1	2	3	4	5	6	7
I find my experience with this website fun	1	2	3	4	5	6	7

3. To what extent do you think that the visual appeal affects your attitude towards product?

Questions	Responses						
	Strongly disagree				Strongly agree		
The product that I' ve just examined is good	1	2	3	4	5	6	7
I have formed a favorable impression towards the product that I' ve just examined	1	2	3	4	5	6	7
I like the product that I' ve just examined	1	2	3	4	5	6	7

4. To what extent do you think that the shopping enjoyment affects your attitude towards online grocery store?

Questions	Responses						
	Strongly disagree						Strongly agree
I like shopping on this online grocery store	1	2	3	4	5	6	7
Shopping on this online grocery store is a good idea	1	2	3	4	5	6	7
I think this grocery store is appealing	1	2	3	4	5	6	7
I think this grocery store is favorable	1	2	3	4	5	6	7

5.To what extent do you think that the attitude towards to product affects your purchase intention?

Questions	Responses						
	Strongly disagree						Strongly agree
It is likely that I will buy this grocery	1	2	3	4	5	6	7
I will definitely try this grocery	1	2	3	4	5	6	7
I have strong intentions to purchase the grocery	1	2	3	4	5	6	7
The probability that I would consider buying grocery from this website is high	1	2	3	4	5	6	7
If I were to buy grocery, I would consider buying it from this store	1	2	3	4	5	6	7

6.To what extent do you think that the privacy concern affects your online shopping?

Questions	Responses						
	Strongly disagree						Strongly agree
I am concerned that the information I submit to Internet could be misused	1	2	3	4	5	6	7
I am concerned that a person can find private information about me on the Internet	1	2	3	4	5	6	7
I am concerned about providing personal information to Internet, because of what others might do with it	1	2	3	4	5	6	7
I am concerned about providing personal information to Internet, because it could be used in a way I did not foresee	1	2	3	4	5	6	7

7. To what extent do you think that the involvement affects the topic of this study?

Questions	Responses						
	Strongly disagree						Strongly agree
I am much involved in the topic of this grocery store	1	2	3	4	5	6	7
Much of the issue discussed in this grocery store would be on my mind lately	1	2	3	4	5	6	7

8. To what extent do you think that your expertise about this topic?

Questions	Responses						
	Strongly disagree						Strongly agree
I am much informed on the subject matter of this issue in this online grocery store	1	2	3	4	5	6	7
I am an expert on the topic of this online grocery store	1	2	3	4	5	6	7

9. Overall, I believe that the given product presented results of this online grocery store are:

Questions	Responses						
	Strongly disagree						Strongly agree
Misleading	1	2	3	4	5	6	7
Deceptive	1	2	3	4	5	6	7
Biased	1	2	3	4	5	6	7

APPENDIX G

生鲜食品网购调查问卷

亲爱的朋友：

您好！欢迎您参加本次调查。我们希望能从您那里了解一些关于网购生鲜食品方面的信息，大概会占用您 10 分钟时间。本问卷无须署名，所有数据仅用于课题研究，不会泄露您的任何个人信息，请您放心、客观地作答。谢谢您的参与！

以下是关于您购买牛排时考虑的因素，请根据您刚才浏览的网页，结合您的认同程度进行打分：1-非常反对，2-反对，3-有些反对，4-中立，5-有些同意，6-同意，7-非常同意。

1. 以上牛排图片的设计多大程度上影响您的视觉吸引？

问题	1	2	3	4	5	6	7
这张图片令人赏心悦目	○	○	○	○	○	○	○
这张图片的设计令人赏心悦目	○	○	○	○	○	○	○
从审美角度看，这张图片很吸引人	○	○	○	○	○	○	○
这张图片展示产品的方式很吸引人	○	○	○	○	○	○	○
我喜欢这张图片展示出来的样子	○	○	○	○	○	○	○

2. 以上牛排图片所在的虚拟网站多大程度上影响您的购物乐趣?

问题	1	2	3	4	5	6	7
体验这个网站很有趣	○	○	○	○	○	○	○
体验这个网站令人愉快	○	○	○	○	○	○	○
体验这个网站令人兴奋	○	○	○	○	○	○	○
体验这个网站令人感觉好玩	○	○	○	○	○	○	○

3. 您对该牛排的态度:

问题	1	2	3	4	5	6	7
刚才看过的牛排不错	○	○	○	○	○	○	○
刚才看过的牛排给我留下了好的印象	○	○	○	○	○	○	○
我喜欢刚才看到的牛排	○	○	○	○	○	○	○

4. 您对这个生鲜网站的态度:

问题	1	2	3	4	5	6	7
我喜欢在这个生鲜网站上购物	○	○	○	○	○	○	○
在这个生鲜网站购物是个好主意	○	○	○	○	○	○	○
在这个生鲜网站购物很吸引人	○	○	○	○	○	○	○
我认为这个生鲜网站受人喜爱	○	○	○	○	○	○	○

5. 您购买这款牛排的意愿:

问题	1	2	3	4	5	6	7
我很可能会买这款牛排	○	○	○	○	○	○	○
我肯定要买这款牛排	○	○	○	○	○	○	○
我买这款牛排的意愿很强烈	○	○	○	○	○	○	○
我很可能会考虑购买这款牛排	○	○	○	○	○	○	○
如果要买牛排,我会考虑在这家网站上购买	○	○	○	○	○	○	○

6. 您对这个网站的忠诚度：

问题	1	2	3	4	5	6	7
每当我要购买牛排的时候，我就尽可能到这个网站	○	○	○	○	○	○	○
每当我要购买牛排的时候，我首选这个网站	○	○	○	○	○	○	○
我喜欢用这个生鲜网站	○	○	○	○	○	○	○
对我来说，这个网站是网上买生鲜的最佳选择	○	○	○	○	○	○	○
我想这是我最喜欢的牛排购物网站	○	○	○	○	○	○	○
下次我要买牛排的时候，我愿意使用这个网站	○	○	○	○	○	○	○

7. 您对个人隐私问题的看法：

问题	1	2	3	4	5	6	7
我担心提交到网上的个人信息可能会被滥用	○	○	○	○	○	○	○
我担心有人能在网上发现我的个人隐私	○	○	○	○	○	○	○
我担心提交到网上的个人信息，别人会利用	○	○	○	○	○	○	○
我担心提交到网上的个人信息，以某种不可预料的方式被利用	○	○	○	○	○	○	○

8. 您对生鲜图片装饰设计这个主题的了解程度：

问题	1	2	3	4	5	6	7
我对生鲜食品图片装饰设计这个相关的主题接触颇多	○	○	○	○	○	○	○
近来，与图片装饰设计相关的很多问题，我都有所思考	○	○	○	○	○	○	○

9. 您对这个主题的专业知识程度：

问题	7	6	5	4	3	2	1
关于生鲜食品图片装饰设计的这个话题，我了解得非常多	○	○	○	○	○	○	○
对于生鲜食品图片装饰设计相关的话题，我有专业的了解	○	○	○	○	○	○	○

10. 总的来说，您认为这家生鲜网站展示的牛排是：

问题	7	6	5	4	3	2	1
误导	○	○	○	○	○	○	○
欺骗	○	○	○	○	○	○	○
有偏差	○	○	○	○	○	○	○

11. 您的性别？

○ 男

○ 女

12. 您的年龄？

○ 20 ~ 35 岁

○ 36 ~ 45 岁

○ 46 ~ 55 岁

○ 56 岁以上

13. 您一般每隔多久会在网上购买生鲜食品？

○ 大约每周一次

○ 大约每个月两三次

○ 大约每个月一次

○ 大约几个月一次

再次感谢您的参与！

APPENDIX H

Some terms are used throughout this study. In order to create an understanding of the fundamental terms associated with this study, specific terms have been defined.

Reflective surface is intentionally made by photographer.Some gradient lighting and close-up technique are adopted to present the products' brightness and lights. In this study, reflective surface represents the symbol of freshness, which can easily make consumers to recognize that the target grocery could be fresh. When customers find the reflective surface of grocery, it could be a cue to make a consumer think that this grocery is fresh.

Color contrast is defined as a change in color appearance of a test stimulus with a change of surround chromaticity (i.e., hue, value, and chroma) (Pereverzeva and Teller, 2009) . Color contrast on webpage is used to reduce consumers' information processing load (Budimir and Palmovic, 2011), to highlight products or information, and to shorten and simplify the path from stimulus to pleasantness. In this study, contrast color usage between the target product and background could easily make customer attend to the target product. Photographer intentionally displays a target product with background in contrast color.

Complementary goods layout refers to the products' location in the photo and how consumers' concern is focused. Combination of the target product and complementary goods could easily make customer recall the way of consumption. Photographer intentionally selects the complementary goods as a background in a picture. Layout is one of photos design artifacts. When groceries are placed in the picture based on some complementary goods as the background, layout may impact consumer's attraction and perception about the target product. It can make consumer easily value them as well as recall the way of consuming them.

Photo decoration cues mean various photographic and editing techniques (i.e. camera control, focusing, lightning, exposure, rendering, and balancing) adopted to polish pictures and they intend to emphasize or hide a part or overall target contents for effective communication by changing brightness, color, graphics and layout.

Visual appeal can be defined as "the tangible aspect of the online environment which reflects the 'look and feel' or perceived attractiveness of a website" (Weiss et al., 2003). Visual appeal connotes the attractiveness of the website, consisting of photos, colors, and fonts (Loiacono, 2003; 2007) which is inspired by the design, beauty inherent and physical attractive appearance in the retail setting (Holbrook, 1994).

Shopping enjoyment, as an affective assessment criterion of the shopping experience, refers to the shopping experience full of enjoyment in its own right, excluding the received product's value (Cai et al., 2009).

Attitude towards product means consumers' overall evaluations of products.

Attitude towards store refers to consumers' general evaluations of a shopping experience towards a specific Web store.